LEADERSHIP FOR AGRICULTURAL INDUSTRY

CAREER PREPARATION FOR AGRICULTURE/AGRIBUSINESS

Max L. Amberson, Consulting Editor
Montana State University

WORKING IN AGRICULTURAL INDUSTRY

Jasper S. Lee
Mississippi State University

WORKING IN PLANT SCIENCE

Douglas D. Bishop
Montana State University

Stephen R. Chapman
Clemson University

Lark P. Carter
Montana State University

WORKING IN ANIMAL SCIENCE

Paul Peterson
California State
Polytechnic University

Allen C. Christensen
California State
Polytechnic University

Edward A. Nelson
California State
Polytechnic University

WORKING IN AGRICULTURAL MECHANICS

Glen C. Shinn
University of Missouri

Curtis R. Weston
University of Missouri

LEADERSHIP FOR AGRICULTURAL INDUSTRY

Bob R. Stewart
University of Missouri

LEARNING THROUGH EXPERIENCE IN AGRICULTURAL INDUSTRY

Max L. Amberson
Montana State University

B. Harold Anderson
Colorado State University

Each instructional module is accompanied by the following items:

- An introductory sound filmstrip with teacher guide notes
- A student textbook
- A student activity guide
- A teacher's manual and key
- A set of transparency masters

A comprehensive Program Planning Guide is also available to assist in planning and implementing a core curriculum in agriculture/agribusiness.

LEADERSHIP FOR AGRICULTURAL INDUSTRY

Bob R. Stewart
University of Missouri
Columbia, Missouri

Max L. Amberson
Consulting Editor
Montana State University
Bozeman, Montana

GREGG DIVISION

McGRAW-HILL BOOK COMPANY

New York
St. Louis
Dallas
San Francisco
Auckland
Bogotá
Düsseldorf
Johannesburg
London
Madrid
Mexico
Montreal
New Delhi
Panama
Paris
São Paulo
Singapore
Sydney
Tokyo
Toronto

Library of Congress Cataloging in Publication Data

Stewart, Bob R
Leadership for agricultural industry.

(Career preparation for agriculture/agribusiness)
Includes index.
1. Agricultural—Vocational guidance. 2. Agricultural industries—Vocational guidance. 3. Leadership.
4. Future Farmers of America. I. Title. II. Series.
S494.5.A4S73 630'.203 77-14406
ISBN 0-07-000847-7

Leadership for Agricultural Industry

1234567890 VHVH 783210987

The editors for this book were Susan Horowitz and Susan Berkowitz, the designer was Sullivan-Keithley, Inc., the art supervisor was George T. Resch, and the production supervisor was S. Steven Canaris. It was set in Times Roman by John C. Meyer & Son. Cover design by Ed Aho. Printed and bound by Von Hoffman Press, Inc.

CONTENTS

PICTURE CREDITS

Page 6: top and bottom left, Future Farmers of America, photograph by Susan Berkowitz; right, Grant Heilman.
Page 8: left, Monkmeyer Press, photograph by Perry Ruben; right, Monkmeyer Press, photograph by Nancy Hays.
Page 11: left, Diane Mark; right, United States Department of State Photography.
Page 17: The Granger Collection.
Page 22: Future Farmers of America, photograph by Susan Berkowitz.
Page 31: top, Future Farmers of America, photograph by Susan Berkowitz; bottom, National 4-H Council.
Page 34: Susan Berkowitz.
Page 35: Future Farmers of America.
Page 37: National 4-H Council.
Page 40: top and bottom, Future Farmers of America, photograph by Susan Berkowitz.
Page 44: left, Cotton Incorporated; right, John Deere.
Page 55: top right and bottom, Future Farmers of America, photograph by Susan Berkowitz; top left, YMCA, photograph by King Photographers.
Page 56: Cincinnati Reds.
Page 60: National 4-H Council.
Page 61: top, Metropolitan Life; bottom, American Telephone and Telegraph.
Page 64: top, United Nations; bottom, National 4-H Council.
Page 65: left, United Nations, photograph by Michael Tzovaras; right, Monkmeyer Press, photograph by Nancy Hays.
Page 69: Future Farmers of America.
Page 71: Future Farmers of America.
Page 74: top and bottom, American Telephone and Telegraph.
Page 78: Monkmeyer Press: top left, Sybil Shackman; bottom left, Mimi Forsyth; right, David Stridder.
Page 80: Future Farmers of America, photograph by Susan Berkowitz.
Page 86: Future Farmers of America.
Page 92: Monkmeyer Press, photograph by Mahon.
Page 96: Monkmeyer Press, photograph by Mimi Forsyth.
Page 98: United Nations.
Page 113: Monkmeyer Press, photograph by Sybil Shackman.
Page 115: Future Farmers of America, photograph by Susan Berkowitz.
Page 121: Monkmeyer Press, photograph by Mimi Forsyth.
Page 129: Future Farmers of America, photograph by Susan Berkowitz.
Page 131: Future Farmers of America.
Page 133: All four photographs, Future Farmers of America.
Page 135: Future Farmers of America.
Page 138: top, Ginger Chih; middle, Future Farmers of America, photograph by Susan Berkowitz; bottom, Future Farmers of America.
Page 142: Future Farmers of America, photograph by Susan Berkowitz.
Page 144: American Society for the Prevention of Cruelty to Animals.

Back cover left: National 4-H Council.
Back cover middle and front cover, Future Farmers of America, photographs by Susan Berkowitz.

PREFACE

You can raise a Holstein cow. You can describe the life cycle of plants. You have welding skills and can construct a livestock feeder. But can you be a leader and work with people?

People working with people: on the job, in the classroom, in youth organizations, in family and social settings. The development of leadership abilities and personal skills associated with working with people is vitally important to students preparing for careers in the diverse and dynamic agricultural industry. Studies have shown that the majority of people who fail on jobs fail not because of a lack of technical skills but because of a lack of leadership and human relations skills.

Agricultural industry has become big business and big business needs leaders.

It is becoming increasingly important for producers on farms and ranches to be able to work with others to maximize their productivity as well as to assume positions of leadership as they relate agriculture to the urban segment of our society. The owner or manager of a large farm also requires abilities in leadership, supervision, and human relations. And the owners, managers, and employees of agribusiness firms must be able to work with others using effective human relations and communication skills.

There is nothing new about the need for leadership and personal development activities in vocational agriculture. In fact, the FFA was developed in response to this need and is based on the concept of fostering leadership, citizenship, and personal growth.

Leadership for Agricultural Industry: An Instructional Module

Leadership for Agricultural Industry was developed as an instructional module to assist teachers in implementing a meaningful unit in leadership and personal development and to prepare students to utilize the FFA as a laboratory for applying leadership skills learned in class. In this module, membership in the FFA is presented as both a desirable and necessary step to develop the leadership skills essential for successful employment and advancement in production agriculture or in agribusiness.

The instructional module *Leadership for Agricultural Industry* includes the following components:

- An introductory sound filmstrip that shows vo-ag students exhibiting leadership skills
- A student textbook that provides basic leadership skills in a career context
- A student activity guide that includes application-type questions, tests, and long- and short-term projects (problem-solving situations) to supplement learning of basic concepts
- A set of overhead transparency masters that facilitates classroom discussion and demonstration
- A teacher's manual and key that includes teaching suggestions, answers, and objective tests to assist in implementation of the instructional module

Student Textbook

Leadership for Agricultural Industry assists students in developing leadership and personal skills by using a competency-based approach and behavioral goals.

The text is divided into units and chapters. An occupational matrix begins each unit. The matrix shows the relationship between jobs in agricultural industry and competencies needed to enter and succeed in those jobs.

On the horizontal axis of the matrix are sample job titles from the seven major career areas of agriculture/agribusiness. The sample jobs are high-frequency jobs that students with vocational agriculture preparation and supervised occupational experience could enter after graduation.

The vertical axis of the matrix shows competencies needed by students who plan to enter the occupations listed. These competencies, which include knowledge, skills, attitudes, and experiences, are designated as *very important*, *important*, or *not important* for the jobs listed. The competencies listed in the occupational matrices also are developed in the chapters of the text.

Goals stated at the beginning of each chapter help students become aware of learning expectations. End-of-chapter questions are designed to aid students in meeting the stated goals of the chapter. Examples, case studies, and illustrations relate general concepts to world-of-work and daily living situations.

Instructional Strategy

The organization of the instructional module was based on competency studies relating to leadership and personal development needs for people working in the field of agriculture/agribusiness and reviews of suggested curriculum units related to FFA and leadership development from several states. The FFA section was developed to complement the use of the *Official FFA Manual* and the *FFA Student Handbook* available from the National FFA Center, Alexandria, Virginia.

The instructional module is organized into three units: (1) "Individual Growth and Leadership," (2) "Working with Others," and (3) "Developing Leadership Abilities." The first unit focuses on leadership and personal development skills and their importance to the individual. Unit II focuses on developing useful skills in working with and influencing others, both individually and in groups. Concepts in this unit include human relations, group dynamics, personal and group communication skills, and parliamentary procedures.

The final unit focuses on ways to develop the leadership and personal skills important to those who plan to work in agricultural industry. The unit deals with the purpose of organizations and the responsibilities of leaders as members and officers. The FFA is emphasized as a leadership laboratory. The module concludes with a focus on leadership skills in the world of work.

Career Preparation for Agriculture/Agribusiness

Leadership for Agricultural Industry is part of an instructional program designed as a core curriculum for students of vocational agriculture. The purpose of the core curriculum is to provide basic principles and entry-level employment skills for agriculture/agribusiness students. The core provides solid foundation for advanced specialization. The titles in the core curriculum are:

- Working in Agricultural Industry
- Working in Plant Science
- Working in Animal Science
- Working in Agricultural Mechanics
- Leadership for Agricultural Industry
- Learning through Experience in Agricultural Industry

A comprehensive program-planning guide has been developed to assist in the implementation of a one- or two-year competency-based core curriculum.

Acknowledgments

Many people assisted in the preparation of this text. However, special recognition should go to Max L. Amberson, Montana State University, Consulting Editor, for constructive comments that contributed to the development of this program, as well as to three vocational agriculture teachers whose manuscript reviews helped strengthen the quality of the material: B. H. Claxton, Jeff Davis High School, Hazlehurst, Georgia; Peter Edgecomb, Limestone Junior-Senior High School, Limestone, Maine; and James J. Hubbell, Assumption High School, Napoleonville, Louisiana.

The teachability—format, approach, reading level—of the program was confirmed through field testing of selected units.

UNIT I

LEADERSHIP AND PERSONAL GROWTH

COMPETENCIES

Competencies	PRODUCTION AGRICULTURE Crop farmer	AGRICULTURAL SUPPLIES/SERVICES Coop manager	AGRICULTURAL MECHANICS Machinery salesperson	AGRICULTURAL PRODUCTS, PROCESSING, AND MARKETING Butcher	HORTICULTURE Nursery supervisor	FORESTRY Forestry aide	RENEWABLE NATURAL RESOURCES Soil conservationist
Demonstrate the ability to lead and to follow	Very Important	Very Important	Very Important	Very Important	Very Important	Very Important	Very Important
Show respect for the rights of others	Very Important	Very Important	Very Important	Very Important	Very Important	Very Important	Very Important
Develop a plan for personal and professional self-improvement	Very Important	Very Important	Very Important	Very Important	Very Important	Very Important	Very Important
Develop the ability to work independently	Very Important	Very Important	Very Important	Very Important	Very Important	Very Important	Very Important
Exhibit a willingness to work	Very Important	Very Important	Very Important	Very Important	Very Important	Very Important	Very Important
Demonstrate acceptable personal appearance and social traits	Important	Very Important	Very Important	Very Important	Important	Important	Very Important
Show personal integrity	Very Important	Very Important	Very Important	Very Important	Very Important	Very Important	Very Important
Budget time effectively	Very Important	Very Important	Very Important	Very Important	Very Important	Very Important	Very Important
Show a willingness to take supervision	Important	Very Important	Very Important	Very Important	Important	Very Important	Very Important
Demonstrate the ability to make decisions	Very Important	Very Important	Very Important	Very Important	Very Important	Very Important	Very Important

Very Important

Important

Not Important

You walk into the kitchen one morning. Your parents are worrying about how they can keep the family farm supply business open late during the busy season and still have time to get ready for the cookout they had been planning for their employees and best customers.

"What can we have to eat for the cookout and how do we find time to do the shopping?" your father wonders.

"Hamburgers," you suggest. "There are plenty in the freezer. And the lawn needs mowing, too."

"That's good thinking," says your father with a smile.

That afternoon, your parents come in from work to find you mowing the lawn. "Hi, I remembered that the tables would have to be set up, too," you greet them.

"Thank you very much," says your mother. "It's good to see you taking on so much responsibility. You're showing real leadership."

You understand about responsibility, but what did she mean by leadership? You were just trying to be helpful. But in giving your parents a much-needed hand, you showed several important leadership characteristics: the ability to make decisions, to take the initiative, to work with others, and to follow through.

These all are important characteristics for you to develop for everyday living. They help you to act responsibly in your family, your school, and in the organizations you have joined. Acting responsibly often leads to and is a part of exercising leadership.

Leadership may involve running a company, governing a state, presiding at a meeting, or leading an army. But it is more than this. Parents are leaders in the family, and teachers are leaders in the classroom. Older children often exercise leadership by taking on responsibility in the family. Bus drivers, flight attendants, police officers—they all have leadership roles.

So what is leadership? How do you know when you are leading? Are *you* a leader or a follower? Are you able to assume leadership? Can you learn to be a leader? You may have asked yourself some of these questions already.

Maybe you don't think of yourself as a leader right now. But you can look at leadership from an individual *or* a group viewpoint. This suggests that there are many ways an individual functions in society.

If you look at the job matrix on the facing page, you will see that many of the abilities needed for leadership also are used every day by people in jobs and in other situations. You can see that you already need or possess a number of these abilities. They are required for success in school, as a member of a religious organization, or as a member of a school organization such as the Future Farmers of America (FFA). They are needed in order to work successfully with other people.

Leadership development and personal growth are important in all aspects of your life, starting right now. Your decisions and actions of today are important, whether they deal with your school work, one of your organizations, or your part-time job. The ideas discussed in this unit will help you develop the abilities that you can use immediately, and for the rest of your life.

CHAPTER 1

LEADERSHIP EXAMINED

Are you a leader or a follower? You fill both roles in your lifetime. You probably have acted as a leader already, maybe without thinking about it. If you have been president of your school class, a Scout leader, an FFA chapter officer, or captain of an athletic team, you obviously were acting as a leader. But you also show leadership when you make suggestions, help other people decide what to do, or make sure that things get done.

> Carla had been in the FFA chapter for 6 months. She did not hold an office or think of herself as a leader. But at one meeting, the members were discussing whether or not to approve the chapter's budget. Carla questioned whether too much money was budgeted for entertainment. This led to a lively discussion and a revised amount. She later was asked to serve on a committee set up to study and propose different fund-raising projects.

Like Carla, leaders work with other people to accomplish common tasks. This makes everybody a leader. You know that the President of the United States is a leader. So is the mayor of a town, a school principal, or a chief of police. But what about *you*? Can you think of times when you acted as a leader?

The simplest act of leadership is the suggestion that one person makes when two or more people are trying to decide what to do. The suggestion may be accepted for many reasons. Maybe the person has good ideas. Or perhaps the person is nice, popular, older, or just louder. Each of these characteristics may affect leadership at one time or another.

> Albert and Judy worked together in a nursery. Their supervisor had told them to make a display of plants on a table in front of the store, but she had not given them specific instructions. Judy suggested that the flowering plants be used as a border. Albert accepted this suggestion—and thus Judy's leadership—because it seemed like a good idea. Later, during lunch break, Albert suggested that they buy some bread, milk, and cheese at the Farmer's Market and have lunch in the park. This time Judy accepted Albert's suggestion and leadership.

Each person needs to learn about leadership and how to develop as a leader. Many of the characteristics of leadership are the same as the ones needed to lead an effective life. So what is leadership, and what does a leader do?

CHAPTER GOALS

In this chapter your goals are:

- To define leadership
- To distinguish between formal and informal leadership
- To analyze the role of a democratic leader
- To identify the personal characteristics of leadership
- To explain why each person can be a leader at some time

Leadership Defined

The word *leadership* can be used in many ways. What does it mean? Basically, it is the act of leading. Most dictionaries define leadership as guiding or conducting others. Leadership involves directing the actions, thoughts, and opinions of others. It also means to instruct or teach. As you begin to focus on the meaning of leadership, you begin to see how you lead.

If members of an FFA chapter from another school were to visit your school, one simple act of leadership would be to show them around your school and community. You would guide them to sites of interest. The purpose of the tour would be to instruct or teach the visitors about the interesting and important things that they would see. You would be directing their actions. One of the stops you might make is in front of a statue in the town square near the courthouse: "Look at the statue over there. That person was responsible for saving the town from disaster during the big flood of 1887."

Leaders can serve as teachers, guides, and organizers.

Many people have studied leadership and leaders. They have used different approaches to try to find out more about leadership. Some have studied leaders themselves to learn about their personalities, their ability to convince others, or their methods of using power. Other persons have examined the ways leaders caused people to change. These studies have found that many factors influence leadership.

Some people can lead because they are well liked. Other leaders have a great ability to convince people to act by talking to them individually, by discussing issues in small groups, or by making speeches to larger groups. Still others use *power*, that is, force or a position of authority. They can *make* people do what they want. But leaders usually have the ability to use a combination of techniques.

Your teachers have authority over, and responsibility for, what happens in the classroom. The most effective teachers lead their classes by being interesting, well liked, organized, and able to talk well. They have a good knowledge of the subjects being discussed and know how to get students to take part in the discussions. They are interested in their students. You probably judge your teachers on the basis of some of these characteristics. If you have not done so before, try to think of your teachers as leaders and figure out which characteristics make them effective.

Types of Leadership

You can observe leadership in organized situations, such as your FFA chapter or your school. Just as important, leadership can also be observed in person-to-person relationships that are not formally organized. You need to be able to understand both situations. It is important for you to learn, as you work with groups, to recognize the difference between formal and informal leadership. Both kinds are necessary for the efficient operation of a group. And each person exercises both types of leadership.

Formal Leadership. *Formal leaders* are people who have been elected or appointed leaders of an organization. The officers and committee chairpersons of the FFA are formal leaders. So is the president of an agricultural supply company or the supervisor in a fertilizer factory. Formal leaders are those who have been given authority and responsibility. Often their powers and duties are spelled out in an organization's bylaws.

Formal leaders are responsible for the functioning of an organization. They must see that activities are planned and carried out. In a democratic organization, they see that the wishes of the members are followed. They are expected to help identify and clarify group goals and to guide the group in meeting its goals.

Informal Leadership. Formal leaders can't do the leadership job by themselves. The President of the United States consults many advisors before making a decision. Some of these advisors may not have formal positions within the government. They may be specialists or business executives whom the president trusts. Such advisors are informal leaders. *Informal leadership* is exerted by persons who have influence within a group, but who do not hold elected or appointed positions.

Informal leaders are important in the functioning of a democratic group. Often their role is of central importance to the success of an organization. Not everybody can be a formal leader, but there is always room for informal leadership. Often, informal leaders *specialize*. They become interested in a topic and exert their leadership when that topic is discussed. If you do a lot of work on raising money for your FFA chapter, you probably will become an informal leader when fund-raising decisions are made. If you become a real expert, you may be as much of a leader in this area as the treasurer and the chairperson of the earnings and savings committee, who have the formal responsibility.

Relationship between Formal and Informal Leaders. For the good of an organization, the relationship between formal and informal leaders needs to be handled with care and tact. Informal leaders sometimes assume leadership or raise opposition on certain questions. Such actions may place informal leaders in conflict with formal leaders. If carried to extremes, such conflicts could harm the group. Informal leaders should be careful to see that their role helps the organization. Airing all sides of a question is a part of the democratic process and one of the functions of informal leadership. But so is compromise.

On the other hand, formal leaders need to be sensitive to who the opinion leaders are in a group. The skilled *formal* leader tries to work *informally* on a person-to-person basis to seek support or to reconcile differences. For example, if the FFA officers know you are an expert in the area of fund raising, they should try to involve you in discussions before decisions are made. (Your informal leadership would be shown by a willingness to cooperate.) In this way, formal leaders can enlist the support of a group's informal leadership.

Both Kinds Are Needed. Jerry Litton was a national FFA officer and a successful leader in the agricultural community. He was then elected to the U.S. House of Representatives as a Democratic representative from Missouri. In his government position, Litton acted as a formal leader. But at other times, he gave advice and worked outside of Congress to support the interests of agriculture. At such times, he was acting as an informal leader.

Organizations need both formal and informal leadership. The formal leaders must get the support and the ideas of the other members if a group is to be active and successful in meeting its goals. Obviously,

The formal leader is elected or appointed. The informal leader has influence without holding a specific position.

the FFA officers cannot make an exciting organization by themselves. But they can help make the organization exciting and productive if they work with the informal leaders and the other members.

Values of Leadership

American democracy is founded on the concept of shared leadership and wide participation by many members of a group in making and carrying out group decisions. This concept is essential to the successful operation of the FFA, the Kiwanis Club, the family, the farm operation, or the local nursery. Leadership abilities are needed by people serving their communities or planning recreational activities just as much as they are needed by people operating businesses or running governments.

A study of leadership will help you get ready to help others. It will also help you gain abilities you need to get a good job. Leadership brings rewards and benefits. People who have developed their leadership ability enjoy the respect of others. People who can lead are rewarded with better-paying jobs and with promotions, but they should also be ready and able to assume greater responsibility. Most promotions bring greater demands for leadership and increased responsibilities. Leaders take pride in helping others and gain a sense of security and self-confidence.

Leaders are not born. They develop themselves through study, hard work, and the ability and willingness to help others. Your development of leadership abilities will depend on how well you develop and use your own abilities in everyday life.

The development of leadership ability is vital to the continued growth of organizations, businesses, communities, and the nation. Responsible group members and citizens work hard to learn to use their abilities to lead.

Styles of Leadership

Leadership styles are generally defined in terms of two distinct positions—autocratic and democratic. However, it is rare to find either style in a pure form.

Autocratic Leadership. *Autocratic leaders* use their power and positions to force their decisions and desires for action on others. Other people are simply expected to do as they are told. They are not asked for any suggestions. Indeed, they may be punished for making suggestions. If you have read *Mutiny on the Bounty* or seen the movie, you can see that Captain Bligh was an autocratic leader. He expected immediate obedience from his crew. His authority came from his position as a captain in the British Navy.

There are many examples of autocratic leaders in world history from the pharaohs of Egypt in ancient times to the Shah of Iran today. Joseph Stalin ruled the Soviet Union autocratically for 30 years, using force, power, and fear to get his own way.

Democratic leaders often have to exert autocratic leadership in time of war. During peacetime, the President of the United States or the Prime Minister of Great Britain must have many of their decisions approved by Congress or Parliament. But during World War II, President Franklin D. Roosevelt and Prime Minister Winston Churchill assumed vast authority to make decisions, often without prior approval from their legislative bodies or citizens. Battles were planned and war production was organized without the participation of the citizens of the two nations.

Democratic Leadership. *Democratic leadership* is based on the participation of the members of a group in making decisions. Each person is recognized as important by the democratic leader, whose job is to help the group get things done. The democratic leader also protects the rights of the minority (the smaller number in the group) while the wishes of the majority (the larger number in the group) are being developed and pursued. If there are procedures to be followed, it is the democratic leader's duty to see that they are followed. People are expected to make suggestions and to help make decisions.

A school board president is a democratic leader. Other board members and citizens expect to be heard and to help make decisions. An FFA president is expected to involve all members in planning and carrying out the activities of the chapter. The U.S. government is based on democratic leadership. Citizens are encouraged to make their wishes known to their elected leaders.

Democratic Beliefs. Most Americans accept certain basic beliefs about leadership, just as they do about other areas in life. The democratic style of leadership is a part of the American heritage. It has been important to the success of American democracy. Most Americans believe that democratic leadership is superior to autocratic leadership. This is because they believe that each individual has something to contribute to the group or to the nation. They believe that each person has a unique value that must be protected and the right to act and speak to protect that value. They believe that the collective judgment of a group is better and wiser than the opinion of any one individual.

These beliefs are not just folklore. Sociologists who study how groups function report that participation in making group decisions increases the morale of individuals in the group. Group members are more likely to accept and help carry out decisions they help make. And, importantly, such democratically evolved decisions tend to be better than autocratically imposed decisions.

Thus, one function of a leader in American society is to help groups work in a democratic way—to ensure that individuals participate in making decisions. And each member of the group also is responsible for taking part in group activities and helping the group meet its goals.

The Personal Approach. Most leaders do not follow a pure form of democratic leadership. Formal leaders develop their own personal approach to leadership. And they may vary their approach within a given style.

President John F. Kennedy's approach to leadership was regarded as *charismatic*. This means that people accepted his leadership because they responded favorably to him as a person. He conveyed a sense of idealism and spoke eloquently. He created excitement in his public appearances. President Lyndon B. Johnson shared many of Kennedy's goals and ideals but used a different leadership approach. He preferred the informal setting of small group discussions for getting things done. He would argue and present large amounts of information to support his position.

Both Kennedy and Johnson were democratic leaders. But their approaches varied considerably. Kennedy often appealed to the public to support his programs. Johnson liked to work with his old friends in Congress to get his programs accepted. Both men used a variety of leadership techniques, but each had his own distinct approach to leadership.

Representative Edith Green from Oregon was a leader who used yet another democratic approach. She was chairwoman of a committee of the U.S. House of Representatives that dealt with education. Her approach was to amass information on both sides of a question. She sought the views of experts and of those who would be affected by the legislation that was being considered. Because she "did

Democratic leaders such as Patsy Mink, Assistant Secretary, Bureau of Ocean and International Environment and Scientific Affairs, exhibit both leadership styles in different situations.

her homework,'' she was able to answer questions during debate. For many years, she was effective in getting education legislation passed by the House.

Combined Leadership Styles. Some leaders act autocratically in one situation and democratically in another. A member of Congress will lean toward the democratic style of leadership in dealing with other members during floor debate. The same member may be much more autocratic in dealing with staff members. The president of a business firm who treats employees autocratically probably will act much more democratically at a board of directors meeting.

All leaders have authority, no matter how they are selected or what style of leadership they adopt. The business executive and the FFA president assumed or were assigned authority. This gives them certain powers, but it does not necessarily mean that either of them will be autocratic. They will either use their authority or assign it to others in given situations. The business executive may appoint an employee to close a sale. Or the FFA president may appoint a committee with the power to choose a site for the annual banquet. This is called *delegating authority*.

It cannot be said that democratic leaders do not use authority. To accept leadership means to accept the responsibility to use authority for the best interests of the group. But the use of authority is tempered by democratic beliefs. These beliefs limit the authority of a democratic leader, even one who might have a tendency to be autocratic. The business executive probably wants to hear several opinions before making a decision about how to spend the firm's money. The floor leader in the legislature knows that all members expect to be heard on bills that are important to them and to the voters back home.

Functions of a Democratic Leader

Since democratic leaders cannot simply order that things be done, they must try to get group members involved. They are responsible for directing the activities of the organization. An effective democratic leader works with the group to accomplish its objectives. The leader cannot do everything.

Democratic leaders must share leadership with others. They need to encourage the growth of informal leadership within their group. Those who help carry out actions on which the group decides are leaders. The democratic leader encourages active member participation in making and carrying out decisions. An effective FFA chapter, school organization, business, or government agency needs members who cooperate and work together. Members have to iron out their differences in discussions and in deciding on a course of action.

In helping this process along, the democratic leader ensures that all opinions are heard, that all sides of an issue are discussed, and that all members are encouraged to express their ideas. If the leader fulfills these functions, then the action taken by the group will reflect the will of the majority.

Things to Encourage. Here is a list of things that a democratic leader should try to encourage in exercising leadership authority:

1. Participation by group members
2. Group cooperation
3. The full expression of opinions
4. The use of accurate and adequate information
5. Careful thinking and evaluation
6. The expression of all possible relevant suggestions
7. The expression of differences of opinion
8. A fair hearing for all sides of an issue
9. Proper behavior by members
10. Progress by the group

Recognizing Leaders

How can you recognize a leader when you see one? Can you name five leaders of your country? Can you name three leaders in your classroom? Are *you* a leader?

These are interesting questions. But why did you pick the people you named? Was it their position, reputation, or personal influence? If you named George Washington, it would be because of his position and reputation as the first President of the United States. Maybe you thought of O. J. Simpson because of his reputation and performance as a football player. A teacher may have come to mind because of position, or a classmate because of social or personal influence.

Would any of your classmates name you as a leader? Why or why not? Maybe you are not an official leader by election or by appointment. But you can still work in your classroom and organizations such as the FFA to develop your personal influence, perhaps as an informal leader.

Situations Influence Leadership Styles

You have seen that personality, the ability to convince others, the use of power, and the use of certain techniques all affect leadership. What leaders do, who they are, and where they are also make a difference. A politician uses different techniques while meeting voters than when arguing a point in committee meetings. In each case, the politician may be trying to convince others. But with the voters, the tendency is to be more general and perhaps to appeal to their self-interest or patriotism. In committee meetings, the attempt will be much more specific. The politician will try to argue the merits of the case in detail.

Situations do differ. The football coach may be terrific as a leader on the football field. The coach may mix careful practice, stirring speeches, and tough demands on the players to bring out the best in the team. This approach would be likely to flop at a social action meeting of a religious organization. Much more attention

Leaders often display five common personal characteristics:
(1) *capacity*—as in thinking up new ideas;
(2) *achievement*—as in learning a musical instrument;
(3) *responsibility*—as in finishing a job;
(4) *participation*—as in working with others;
(5) *status*—as in how others view the leader.

would have to be paid to encouraging members than to driving them.

> Mona was captain of her basketball team. She was responsible for calling the plays and was used to having the players follow her directions. But when she was elected class president, she found that the different situation called for different skills. She had to work democratically and to encourage class members to participate in planning and carrying out class projects.

Wise and effective leaders do not rely on personality and power alone. They work and study to develop desirable leadership abilities. They learn how to persuade others by talking and writing and through the example of their own actions. They study the appropriate techniques to influence group actions. Leadership can be learned through reading, through group discussions, by watching leaders in action, and by actually leading. The leader who acquires a large number of leadership skills can use different abilities to meet the needs of different situations.

Personal Characteristics of Leaders

Ralph M. Stogdill, a sociologist, reviewed many studies to try to identify common personal *characteristics*, or traits, of people identified as leaders. Many common traits were identified in the different studies of leaders. However, because the situations and the abilities for leadership varied, not all characteristics were associated with each leader. Nevertheless, a listing of the common traits can be helpful in a study of leadership. This list of characteristics was organized into five areas.

Capacity. This is both the ability to make others feel that you know what needs to be done and the ability to get things done. You can convince others to help you get the job completed. Capacity enables you to communicate ideas and enthusiasm through speaking and acting. Capacity includes characteristics related to the ability to:

- Suggest new ideas
- Make good decisions
- Be cheerful
- Understand and learn
- Size up situations quickly
- Express yourself easily

> Lacey noted that there were hard feelings between the president of the FFA chapter and one of the informal chapter leaders. She suggested that the president invite the informal leader to discuss the supervised occupational experience activity, an area he was interested in. She urged the informal leader to make more of a contribution to the chapter. Lacey's alertness enabled her to size up the situation, and through the use of her verbal ability, she was able to get the two working together.

Achievement. This area concerns present and past accomplishments. You may have a good athletic or scholastic record. To become a leader, you need to relate your achievement to, for example, the tasks of your FFA chapter. Don't rest on your laurels. If you are good in math, you could volunteer to serve on the committee that deals with financial matters. Achievement in English could lead you to write reports. The characteristics of achievement relate to:

- Being good at sports
- Being a good student and a good learner
- Providing the information needed for a job

Responsibility. You come to school in the morning. The desks were disarranged by a meeting that was held the night before. You put them back in order. You have shown *responsibility*, one of the most important areas in leadership and life. It means you can take on and carry out a job for the good of the group. You can be depended on and you keep your word. The characteristics of responsibility include:

- Being willing to work hard
- Being dependable
- Taking initiative
- Completing a job

Participation. You attend every FFA meeting. You help others do the work that needs doing. You keep smiling, even when things are hard, and try to do your best. You adjust quickly to new situations. When a state representative attends a chapter meeting, you introduce yourself and then introduce the visitor to others. The key to participation is involvement. It includes the characteristics of:

- Friendliness
- Ability to work with others
- Humor
- A desire to excel
- Involvement in events
- Easy adjustment to new situations
- Self-confidence

Status. *Status* is a reflection of how others see you. It is determined by your position within a group or in society. Circumstances influence status at the beginning, but status can be changed. You may not be good at sports, but if you work to improve your skills, you can increase your status. You can increase your popularity by being friendly and showing an interest in others. Of course, status is related to the other characteristics of leaders. If you can work to improve them, improvement in your status will probably also occur.

Leadership: A Review

Leadership means to guide, to direct, to instruct. It also means helping others and taking responsibility. While some people are naturally gifted for leadership, the skills can be learned and developed with effort. Some leaders are autocratic; they want to force others to do their will. Most Americans believe that democratic, shared leadership works best.

Certain characteristics have been associated with leadership. Leaders show different characteristics in different situations. They have learned which characteristics are important and work to develop them. But these characteristics are just as important in people's private lives. Everyone must learn to be responsible, to take the initiative, to participate, and to try to achieve in school work, in organizations, and on the job.

THINKING IT THROUGH

1. List five names of people who are leaders.
2. When have you acted as a leader? Explain.
3. How do styles of leadership differ?
4. What are the characteristics of a democratic leader?
5. How do situations influence the actions of a leader?
6. What personal characteristics are important to leaders?

CHAPTER 2

THE INDIVIDUAL AND LEADERSHIP

The nominating committee of your FFA chapter has asked if it can submit your name as a candidate for president. Will you agree to run? Before deciding, you will want to ask yourself a number of questions. How much time will it take? Will it be all your responsibility, or will others help you? Are you good at doing the kinds of things that will be required? Do you like doing them?

You could probably ask dozens of questions. But you will be able to answer all of the questions pretty easily if you have answered one basic question: *Who am I*? The only person who can lead others successfully is the one who has learned to lead her or his own life successfully. To do that you have to know yourself pretty well—your abilities, your talents, your interests, your strengths. And also your weaknesses, your dislikes, and your fears.

If you prefer a casual, informal approach, you probably will not enjoy the FFA presidency. But if you are organized and tidy and feel comfortable in group situations, you probably will say yes to the nominating committee.

But pause a moment. Maybe you lack the abilities and attitudes needed to be FFA president. Yet you think you might like the job. Now is the time to think about it. You can develop and change yourself. You can work to acquire the characteristics you need for leadership and to eliminate the weaknesses that may stand in your way.

And you need not limit yourself to one type of activity. There are many leadership goals. No one personality type is clearly the "leadership" type. Almost everyone has the ability to be a leader in some kind of group or activity. Maybe you can't be an elected officer of the FFA chapter right away. But maybe you can apply specific skills that you have right now to working on a committee or to seeing that a project is carried out. Leadership is needed in all aspects of organizational work.

Different situations call for leadership, and any one situation may require several different kinds of leaders. To find out what kind of leader you want to be and can be, you have to follow that advice from ancient Greece: "Know thyself."

CHAPTER GOALS

In this chapter your goals are:

- To identify the ideals that are important in directing your own life

- To analyze the personal benefits and liabilities of leadership
- To identify the ways an individual can become involved as a leader
- To develop your own list of personal leadership goals

Directing Your Life

Before you can lead others, you must be able to lead yourself. You must be able to handle the major events and decisions in your own life. Certain concepts are important as you move toward directing your own life.

Being Honest with Yourself

Every leadership assignment requires certain abilities of the leader—certain strengths and commitments. You need to be honest with yourself when deciding whether to accept an office in your FFA chapter or when making up your mind about the kind of job or education you will need after high school. If you are honest with yourself, you are more likely to make the right decisions.

> When Marlin was asked to head the committee to plan activities for the FFA chapter, he accepted immediately. He was named because he was good at generating ideas. But Marlin soon learned that leadership meant more than thinking up new ideas. It also meant the capacity to involve others and to follow through on things. He found the assignment a real struggle. Marlin admitted to the FFA advisor and to the committee that he was having problems as chairperson. The advisor encouraged him to learn from the mistake. He advised Marlin to figure out areas where he could provide leadership and to work to improve the weaknesses that made the committee work hard for him. The other committee members pitched in and helped get the work done properly.

Often it is easier to say yes than to say no. Most people want to be agreeable and to please others. Often you may agree to do things without thinking thoroughly about

George Washington Carver—a great American agricultural leader and educator.

what is involved. Or you may want the prestige of leadership. Or maybe you don't like to face your limitations. But sometimes self-honesty may also require that you say yes. Saying yes may be the first step toward overcoming limitations that get in the way of your development.

Angelina was a good writer. She worked hard to develop her ability to write. But she felt shy about working and speaking in groups. She wanted to work on the FFA newsletter, but she was afraid to speak up. Her shyness was getting in the way of her ability to make a contribution. One day she pulled herself together and volunteered for a writing job that had to be done. Her work with others on the project gave her a chance to work to change her shyness.

Weighing the Alternatives

You come upon a live power line that is down. It is lying across the road. What must you do? You must quickly weigh the alternatives. Is it more important to stay and warn others or to go and call for help? Emergencies call for snap decisions. In other situations, you can be more deliberate, but you still have to look at all sides of a question or situation.

Effectiveness in decision making requires that you examine all questions from several different points of view. Think things through and consider the alternatives. Only then are you likely to reach decisions and take actions that are the best ones possible in any given situation.

Nathan was head of an FFA committee that was supposed to decide on projects for the coming year. Lloyd, who had attended only a few meetings, showed up and argued in favor of setting up a small engine-repair shop. He presented many good arguments in favor of the project. Nathan decided to recommend the project to the chapter, even though other committee members had remained silent on the idea. At the meeting of the chapter, the treasurer said there was not enough money to do the project. Many members of the chapter said they did not want to do the project. Nathan had not tried to get a number of opinions or to weigh all sides of the question. The chapter acted wisely in not accepting the recommendation.

Having the Courage of Your Convictions

Every leader needs to be able to reach sound conclusions and to act on them—to make decisions. You cannot constantly look to the group for guidance. If you need the group behind you 100 percent every time, you have surrendered the reins of leadership and become a follower.

President Jimmy Carter had a tough decision to make. He was concerned about the young men who had left this country because they didn't want to fight in the Vietnam War. President Carter felt that these men should be allowed back into the country now that the war was over. Many people shared President Carter's feelings. Many other people disagreed with the President.

Based on his personal conviction, President Carter took action and granted draft evaders *amnesty*, a pardon that allowed them to return to the United States. He was both praised and criticized for his decision. The important thing is that he made a decision. He had the courage as a leader to act in the way he felt was right.

Maybe being a leader will mean you have to say no to members of your group. Maybe someone wants your group to take on another activity. You know that the budget will not support the activity. To say yes in this situation would hurt the organization and the members who wanted the activity.

Following Through on Decisions

The word *decision* suggests the end of a process. You consider a problem from all points of view. You hear what everybody has to say about it. And then you take the final step: you reach a decision. But while this is the end of one process, it is the beginning of another: translating the decision into action.

The successful person not only arrives at wise decisions, but also follows through on those decisions effectively. If you make a decision, you need to put it into action. To make a decision and not to follow through with it is no better, and may be worse than not making a decision at all.

Jane Addams was known for her executive and organizational ability; she knew the value of following through. She wanted to do something for the poor people of Chicago. She decided to create Hull House, a home for poor women. And for the next 40 years she poured her energy into following through on that decision.

You may have noticed that these four ideas are pretty useful for everyone, not just for leaders, in directing their lives. Since you can't lead others effectively until you can lead your own life, you might want to consider how well you measure up against these four ideas: How honest are you with yourself? To what degree do you have the courage of your convictions? Do you thoroughly weigh all alternatives when making decisions? How effectively do you follow through on your decisions?

Leadership and You

At this point, you may be asking yourself: "Do I really want to become a leader?"

Leadership offers both benefits and liabilities which should be weighed carefully by the potential leader.

It's true, leadership *is* demanding. There's more to it than winning a popularity contest. Leadership demands time, energy, and commitment. The essence of leadership is giving to others, although it does have its rewards. But it isn't always a bed of roses. What's in it for you?

Benefits of Leadership

Leadership offers many benefits. Some are *self-centered*: they benefit you as an individual. Others are more *altruistic*: they benefit others. Let's consider each kind.

Benefits to You. One obvious personal benefit of leadership is *status*—the admiration of others. Leadership also brings the respect of others. Status comes with the job, from simply being a leader. But the respect must be earned. It comes from how well you fulfill your leadership role. You also develop your own self-respect as a leader. You learn to value yourself, to appreciate yourself, to like yourself.

Leadership and personal development also bring tangible rewards. One important benefit is employment. Evidence of self-improvement and leadership activities may encourage an employer to hire you. And leadership in certain organizations and occupations results in advancement and higher pay. The more successful you are as a leader, the higher up the ladder you can go, as when a worker becomes a supervisor, then a division head, and then a vice president.

Benefits to Others. The altruistic benefits of leadership include the satisfaction of being of service to others and the satisfaction of making a contribution to the community in which you live. There are positive accomplishments that leaders can achieve that will make the world, or at least a small part of it, a better place to live. There is the reward of a job well done.

Leadership opportunities rarely offer only one benefit. When deciding whether to accept a leadership challenge, it's a good idea to think about all the possible benefits. There is nothing wrong with realizing that some benefits are self-centered, but be wary of saying yes if you're the only one who is likely to benefit. Leadership involves groups of people. Unless your leadership provides some benefits for others, the needs of the group will not be met and your leadership will fail.

Liabilities of Leadership

While you may enjoy many benefits as a leader, you will also discover some liabilities. To be a leader you will need to give some time, hard work, and energy. Your time won't always be your own. There are always the demands of an unexpected crisis or a special situation you did not plan for.

Leadership also carries the risks of failure and criticism. You won't necessarily be praised for the successes of your group, but you will probably be blamed for the failures. And even when you succeed, some of your members may be disappointed and critical because they did not get their way.

And a leader must accept responsibility not only for his or her own actions, but also for the actions of the group as a whole. This isn't always easy. The group may vote to do something that the leader does not support. But the leader must still work for the good of the organization.

Finally, there are always frustrations in working with people. If your tolerance for frustration is low, you will have a hard time trying to lead. Leadership calls for tact and diplomacy and patience. You can't afford to blow your top. Everyone in a group has ideas about how things should be done. Ideas of your own may get lost in the shuffle.

You need to balance the benefits against the liabilities when thinking about taking on a leadership position. But do not dwell too heavily on the burdens of leadership. The challenges usually make the liabilities acceptable. If you are deterred by the liabilities, you may spend your life stuck in situations that offer you no challenge and little personal satisfaction. For example, the supervisor in a farm-machinery repair operation has to think about the records of employees, the inventory of parts, and relations with customers. But there are the rewards of higher pay, doing an important job well, helping train new workers, and serving the customers.

Being of Service

By definition, a leader must meet the needs of other people. You can't lead without followers. Robinson Crusoe on his island may have been an inventor, a philosopher, a farmer, a writer, a hunter, a cook, and an explorer. But he wasn't a leader, at least until Friday showed up. Meeting the needs of others, one of the leader's main jobs, means being of service to them.

To be of service to group members, you must identify their needs first. Then you plan ways to meet those needs. Finally, you must provide the direction and support necessary to help the group meet its needs.

Suppose you are president of a club that wants to provide services to the community. That purpose identifies the members' need: to serve others. As president, you must help the members come up with ideas for meeting that need—ideas about identifying worthy projects, creating a program, or making money. Finally, you must help the club organize its activities in such a way that those plans are carried out.

If, as a result of your efforts, the members do in fact provide a service such as building a park for the community, your presidency—your leadership—will have been a success. You will have been of service to the club and to the community as well.

As a leader, you can serve yourself too. What are your own personal goals and needs? Can any of them be met through assuming leadership in some activity? Will it help you to develop so you can get a better-paying job? So long as your leadership meets the needs of others, there's nothing wrong with its helping meet your needs too.

This could happen in a simple way. Maybe you want to improve your public speaking ability. You could fulfill this need by making reports to your club on the progress of the community service project and by making speeches in the community to drum up support for the project.

A Word of Caution. Not all those who help others meet needs are good leaders. The needs to be served must be legal and worthwhile. To help a gang of young people create a riot to meet their need for attention is not leadership in the sense discussed here. Leadership, as discussed in this book, implies that the activities must have some positive value.

History is littered with leaders such as Hitler, who used his considerable skills to try to murder all Jewish people and to plunge the world into World War II. Al Capone was a leader of a gang of criminals in Chicago. Hitler and Capone served the needs of their followers, in some sense, but those needs were not legitimate. It is up to you to determine the direction of your leadership activities.

Developing as a Leader

Society has a great need for leaders. Federal, state, and local governments need many people to serve in leadership positions. Most jobs also require leadership abilities. In addition, leaders are needed by the many civic and community organizations that provide important services to the nation and the community. Tomorrow's leaders are today's students. Individuals like you must be ready, willing, and able to lead.

Making a Start

Developing your leadership abilities is not especially hard. It takes time and effort, but you can do it if you want to do it. The trick is to pick one or two leadership qualities you want to strengthen in yourself to work on first. Maybe a leader you admire asks a lot of questions to get people involved. With some thought and planning, you can learn to be a questioner.

The nice thing about working on leadership skills is that you don't have to set aside a special time to do it. It's not like a homework assignment. You will find many opportunities to practice leadership skills at home, in school, in clubs or athletic teams, at church, or in the community.

Suppose the first skill you want to develop is the ability to express yourself easily. At home, you can make a point of regularly telling a member of your family about the most interesting thing that happens to you each day. At school, you can discipline yourself to make a spoken contribution to each class each day. Or you can volunteer to present an oral report on a topic your class is studying. You can determine to express your opinion on some aspect of FFA business during every chapter meeting.

It may be hard going at first. But if you really want to improve, you soon will be expressing your ideas more easily. If you don't already belong to a school organization, now is the time to join one. You probably won't have a better opportunity for developing leadership skills than working with an organization like the FFA.

Your challenge is to develop your leadership skills by identifying the personal characteristics you want to improve, to study and work to improve your abilities, and to practice your new-found skills by taking part in activities at home, in school, and in organizations.

Taking a Personal Inventory

A simple way to decide which characteristics you want to improve in yourself is to take an inventory of personal attributes and abilities. This may sound technical, but it is quite simple. Review the personal characteristics of leaders described in Chapter 1. By studying these characteristics, and by being honest and fair as you do, you can develop a clear "road map" to self-improvement. Rate yourself as strong or weak on each characteristic. Don't take

A student compiling a personal inventory sheet.

this as a test of your worth as a human being. It merely shows those areas where you are strong and others where you need improvement. Pick one of the items on which you scored yourself as weak and make that your first target for development.

Personal Inventory Sheet. Another approach to deciding which characteristics you want to improve is to develop a personal inventory sheet. You can list all of your likes and dislikes, your abilities and weaknesses. Make sure you include your strong and weak points. Under strong points you might include friendly, courteous, hard-working, enjoy working with my hands. Under weak points, you might list difficulty expressing myself, hard time with math, often late.

Setting Goals

To a large extent, your future life will be determined by the goals you set for yourself now. Who you will become depends a great deal on who you are right now. Where you will work, who you will work with, where you will live, and even how much money you will make are some of the things that may be affected by decisions and choices that you are in the process of making right now.

Not only must you choose and decide wisely, you must also try to change those things in yourself that may get in the way of becoming the person you hope to become. Most of your life is entirely in your control. Leaders are made, not born. You can change if you want to. You can make plans right now for the rest of your life.

If you have noted your personal strengths and limitations, you already have taken a big step in your planning. Now you can plan your personal goals. Divide these into short-range goals, which can be achieved by the end of the school year, and long-range goals, which may take years, or even a lifetime, to achieve. Then choose one or two goals to work on.

A short-term goal could be to learn parliamentary procedure. A study of a manual or *Roberts' Rules of Order* will teach you the proper procedures for presiding at or taking part in a meeting. Then you can practice these skills in your FFA chapter.

Making personal changes is a slow process. But it's not possible to make any change without beginning to work for that change. Be patient. Don't expect miracles overnight. And, most of all, don't give up if you run into an occasional frustration or setback.

If you have ever tried to learn how to swim, you know how discouraging it can be when you hit a *plateau*—a period when, no matter how hard you try, you can't get your feet off the bottom of the pool. But you also know that if you hang in there and keep trying, sooner or later you will be swimming like a fish. Changing your personality is very much like learning to swim: it's sometimes hard, but you can do it.

Showing Concern for Others

You may have noticed that a number of the personal characteristics of leaders pertain to working with other people. While some leadership qualities, such as ambition or perseverance, have to do with you alone, others are concerned with the people you interact with. You must be able to relate well to other people, to work effectively with others, and to be sensitive to the needs and feelings of others. In living, you are constantly involved with others as an individual or as a leader.

Most jobs bring you into contact with other people. You will have to get along with your coworkers. You also may have

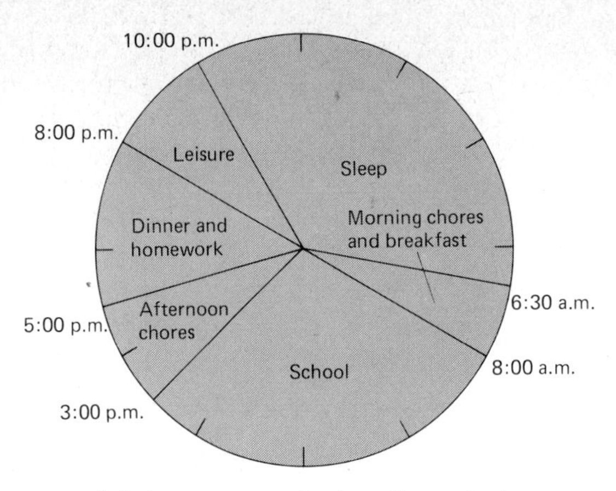

It is important to budget time wisely.

to relate well to the general public. More than half the jobs in our society are service jobs. They provide services, instead of products, to people. So there is more than a 50 to 50 chance that you will be dealing with the public, as well as your coworkers, in any job you take.

Employers want people who can get along with others. Regardless of how much you know, if you can't get along with others, you probably won't be able to keep a job. The ability to get along will determine your friendships, your nonprofessional associations, and your family relationships.

Relationships can be cultivated, much like plants in a greenhouse. And improving your ability to cultivate personal relationships—your willingness to show concern for other people—should be high on your list of personal characteristics that you want to improve.

Planning to Use Time Wisely

"Time is money." This is a favorite expression of busy people. It means that time is something you have to spend in order to do and get what you want. An important personal quality is the ability to budget time wisely and well. Trying to make effective use of your time is worth your best efforts.

For you, time for work at school usually comes first. That's as it should be. School is your work as a student. But how do you use your *leisure time*—that time when you are not working or resting? Leisure time is the time that you can control and use as you wish. You will want to spend some of it on personal recreation and some on helping others.

Make a time budget. Figure out what you do during each period of the day. After you have a picture of your day, you may be able to readjust your schedule to give yourself time to develop your personal and leadership abilities.

Learning to Lead: A Review

You need the same characteristics and abilities in your work and family life as you do to be a leader. You can develop these characteristics and abilities by choosing the ones in which you feel weakest and working to improve them. As you develop your own abilities, you will want to look for opportunities to practice leadership. You will find that it has its rewards and its drawbacks. But by facing the challenges instead of bowing under the burdens, you will find that leadership offers you a chance to grow and develop.

To begin and carry through this process of development requires a great deal of self-knowledge and self-honesty. But if you start to examine your attitudes, abilities, and aptitudes, you will be prepared to handle each situation as it comes along, whether that situation involves leadership or dealing effectively with your own life.

THINKING IT THROUGH

1. How do you exercise leadership in your own life?
2. What are the benefits of being a leader?
3. What are the liabilities of being a leader?
4. What are the advantages of serving others?
5. How can leadership abilities be developed?
6. Why take a personal inventory?
7. How are personal goals set?
8. Why is it important to get along with others?
9. How can time be used wisely?

CHAPTER 3

DEVELOPING AS A PERSON

You must develop first as a confident, responsible person before you can become a confident, responsible leader. Abraham Lincoln was effective as a leader during the Civil War because he had developed as a person by building on his strengths and learning from his mistakes. In his early life, he worked hard to develop his integrity, initiative, perseverance, and ability to communicate effectively. With the strong development of these positive qualities, Lincoln was able to survive two failures in business and several defeats for political office. This combination of strength and knowledge of failure and frustration enabled Lincoln to deal with the failures and frustrations of a divided nation.

Modern leaders like Governor Ella Grasso of Connecticut and Representative Shirley Chisholm of New York worked hard early in life to develop the characteristics that later helped them become effective leaders. Chisholm, for example, took the initiative to learn Spanish so she could better serve the Spanish-speaking people who lived in her neighborhood. Nobody told her she had to do it.

Leaders are made, not born. And they are mostly made by their own efforts, although they may receive help along the way. They work to strengthen their positive characteristics and to improve their weaknesses. In this way, they take responsibility for who they are and what they do.

Your personal characteristics are not fixed forever. They are shaped by the experiences and the environment you have had up to now. They can be changed by new experiences and your reaction to them. To change, you must recognize what you want to change and then seek the new experiences that will enable you to change. Only *you* can make the decision to work to strengthen your positive points and improve your weak areas.

Robert and Carlo were friends who attended the same ninth-grade class. They were, like everybody else, a combination of strong and weak points. Carlo was not well liked. He talked easily, but he liked to argue, tried to borrow homework to copy, and didn't pay attention in class. Robert, on the other hand, was liked, though he was shy, would not speak up in class, and refused to ask for help. However, he did prepare his own lessons correctly and without using other students' work.

Both students recognized that they had problems, and both wanted to change. Carlo determined to turn his negative image around. He tried not to dominate every conversation and began to discuss things rather than to argue. He started to do his own work. He soon noticed that people seemed to like him better. Robert began to change by volunteering to talk in class and to answer questions. This was hard at first, but with practice it became easier. He also began to ask the teacher for help rather than going home frustrated and upset.

Both continued to build on their strengths. Carlo used his verbal ability to begin taking part in class discussions and making oral reports. Robert agreed to prepare a number of written reports. By working to strengthen their positive characteristics while they were trying to improve their weak points, the two classmates built the self-confidence they needed to get them through some tough times in their development.

CHAPTER GOALS

In this chapter your goals are:

- To develop a list of desirable personal characteristics
- To prepare a list of the personal characteristics you want to improve
- To identify important social skills and to develop a plan to practice these skills
- To develop a plan to practice good citizenship

Personal Characteristics

Each person is a mixture of characteristics. Some help a person get along in life; others hold a person back. Laziness is a characteristic most people want to change. Honesty is one that most want to develop and improve. Do your own personal characteristics help or hinder you in getting along with others? Do you attend FFA meetings on time? Do you keep your room neat and clean? Do you complete school work on time? Do you find and do work that needs to be done at home, school, or on the job? Do you attend school regularly? If you answer yes to these questions, you have positive personal characteristics.

Employers look for several characteristics in hiring and promoting people. They are the same ones that people look for and respect in those they work and play with or follow in any situation. These characteristics are as valuable in school or the FFA as they are on the job. Remember that cooperation is the key to making the five I's—*i*nitiative, *i*nvolvement, *i*ntegrity, *i*mage, and *i*mprovement—work for you.

Taking the Initiative

Lisa and Roger wanted to work in politics during an election. Roger went to the headquarters of his favorite candidate and volunteered to help write position papers and to plan strategy. The campaign manager said he did not need that kind of help. "We really need someone to drive workers to political meetings," he said. Roger declined and asked to be called when his skills were needed. He was never called.

Lisa also wanted to draft papers. But when she walked into the headquarters, she noticed a table piled high with envelopes and brochures. Lisa approached the campaign worker who was stuffing brochures in the envelopes. "You look like you could use some help," she said. The worker smiled, nodded, and moved over to give Lisa room to work. Later, Lisa opened mail, answered the phone, did errands, and drove the campaign vehicle. The campaign manager noticed that Lisa always looked for and did the job that needed doing. She was invited to accom-

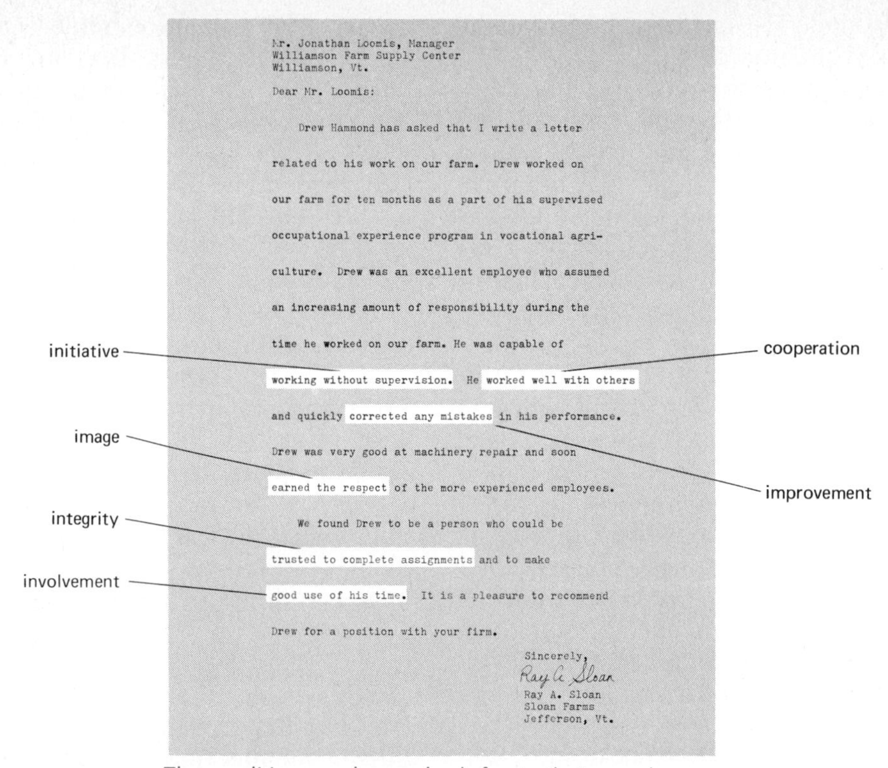
Mr. Jonathan Loomis, Manager
Williamson Farm Supply Center
Williamson, Vt.

Dear Mr. Loomis:

Drew Hammond has asked that I write a letter related to his work on our farm. Drew worked on our farm for ten months as a part of his supervised occupational experience program in vocational agriculture. Drew was an excellent employee who assumed an increasing amount of responsibility during the time he worked on our farm. He was capable of working without supervision. He worked well with others and quickly corrected any mistakes in his performance. Drew was very good at machinery repair and soon earned the respect of the more experienced employees.

We found Drew to be a person who could be trusted to complete assignments and to make good use of his time. It is a pleasure to recommend Drew for a position with your firm.

Sincerely,
Ray A. Sloan
Ray A. Sloan
Sloan Farms
Jefferson, Vt.

The qualities employers look for in their employees.

pany the candidate on the last campaign swing and to take part in strategy sessions.

Lisa showed *initiative*, the ability to start an action and keep it going. She relied on her own resources and had the confidence that she could identify jobs that needed doing. She never did anything for the sake of looking busy. She did not need constant supervision. And by pitching in on some of the seemingly less important jobs, she earned the respect of her coworkers and superiors and the right to do jobs with greater responsibility.

Employers want workers with initiative—people who do not just stand around, but find out what needs to be done without constant supervision. If you had a job in a lawn supply center as a salesperson, you would want to learn your job quickly so your supervisor did not have to instruct you repeatedly. In addition, if you had initiative, you would notice when the shelves needed stocking or straightening or when the floor needed sweeping. You would do these tasks without being told.

You can practice initiative even if you don't have a job right now. You could do needed chores at home without being asked by your parents. You could learn something useful outside of school. In the FFA, you could suggest a new project.

Involvement: The Use of Time

A favorite cartoon character is the person who hangs around the office water cooler all day, except when the supervisor is present. This person is *on* the job, but is not involved *in* the job. This person probably would talk too much to other workers, keeping them from their jobs as well. Time, like money, has value and must be managed to be used productively. Using your time productively to get the most done is what *involvement* means.

Starting *now* to plan your time productively is important. Job applications often ask for a list of your activities. In this way, the employer can get an idea of how you plan your time. Even if you have no prior work experience, these activities will show that you are a person who wants to be involved and who knows how to manage time.

As you become a member of an organization like the FFA, one of the first questions you must ask yourself is how involved you will be. You will have to manage your time. Maybe you could make a list or schedule of how you use your time. How much time do you use for meals, for family chores, for recreation? How much for schoolwork?

After you have figured out how you spend your time, you will be able to rearrange it. How much time will FFA require? There will be meetings to attend and projects to work on. Maybe you have been assigned to a committee. Where will you find time for the added activities? You can give up some TV time. Perhaps the family chores could be done more quickly. Such steps will allow you to become more involved with the FFA activities.

Integrity: A Matter of Trust

Abe and Ellen worked for a farm supply store. One day their supervisor asked for volunteers to reorganize the stockroom. Abe and Ellen said they would do it. They agreed to split up the task, each taking half the area. Abe did his job quickly, using his break and a few hours after work. Ellen went shopping during her break and to a movie after work. She figured she had plenty of time to finish the job. But the stock shipment arrived with her space still not in order. Everyone in the store had to pitch in and do the work she had promised to do.

Ellen might have been surprised if someone had told her she lacked integrity. Like many people, she might have thought that integrity only meant honesty in dealing with money. But *integrity* means you can be trusted and depended on. It means you are honest in all your dealings, whether money is involved or not.

Integrity implies moral soundness. A public official with integrity will not take a bribe or appoint a relative to an important position. If you accept election as a regional representative of your FFA chapter, you promise, in effect, to attend regional meetings and to conduct yourself in a proper manner. Not doing so would hurt your chapter. People would regard you as someone without integrity.

Employers like to work with people they can trust, people with integrity. Employers cannot watch employees every minute to see that they do not steal time, money, or property. They want and have to trust their employees.

Image: How Others See You

Joanne and Luke worked in an agricultural sales office. People called in frequently to place orders. Luke regarded the telephone as an intrusion on his other work. He was gruff and cut people off. Joanne was always polite and helpful. She would transfer calls to other telephones and take messages. She displayed a positive attitude and projected a positive image of herself and the company. Luke's voice and attitude projected a negative image.

The way others see you is called your *image*. Your image reflects your *attitudes*, your ways of relating to life. People read attitudes fairly easily, though they may not be aware of doing so. A person with a positive image is easy and fun to be around. Such a person makes you feel comfortable. A person with a negative image is unpleasant to be around. Nobody likes the person who is grumpy and angry all the time. What are the images of some people you know?

How about your own image? Do you try to get along with others? Are you kind, considerate, polite, understanding? Do you treat others as you want them to treat you? Or are you grumpy and sarcastic? Examine your attitudes. Figure out what people like and dislike about you. They probably are the same things that you like and dislike about yourself. Your attitudes are a clue to your image.

Employers like to hire people with positive attitudes and a positive image. At work, you can develop a positive image by being on time, following directions, and assuming the initiative. You can constantly try to improve your work skills. If you are on the sales staff, you can watch experienced salespeople, read books, and ask others what it takes to be a good salesperson.

Improvement: Profiting from Mistakes

Lucille and Helen were new employees in a nursery. Both of them pitched in and tried to do jobs without being told. One day Lucille rearranged the displays and Helen repotted many of the plants. The supervisor praised both of them for their initiative. The supervisor suggested that Lucille put the same kinds of plants together next time and urged Helen to clean up the materials left during the repotting.

Lucille had worked hard and resented the criticism. She rearranged the plants the way she wanted to the next time, ignoring her supervisor's suggestion. After she did this several times, she was told the nursery did not need her services any longer. Helen listened to the supervisor and made sure she cleaned up after herself. She accepted other criticism and worked hard to improve her work. She was promoted and given a raise.

Helen had a basic quality needed for *improvement*. She accepted criticism and used it to improve her work. Lucille could not stand criticism. Helen built on her strong points of initiative and involvement in her work. And she had a positive attitude toward mistakes, viewing them as an opportunity to learn how to do a better job.

Criticism can be positive or negative. A critic is simply one who evaluates. The supervisor's praise of initiative and suggestions for improvement were both forms of criticism. If you want to improve, you will have to learn to evaluate yourself and accept evaluation from others. Accepting criticism is a key to development and a sign of maturity. It often makes the difference between doing better and remaining unchanged.

As a member or leader of an organization, you should listen to what others have to say, analyze what has been said and done, and learn from past actions. Maybe you planned an FFA fund-raising event too near to Christmas. People had spent their money on gifts and did not have much left over for your event. In the future, you will plan fund raisers at a time when people are likely to have money. You made a mistake and can learn from it.

Try to understand people's criticism of you. You need not accept all of it, only what is true. Everyone makes mistakes. The responsible person admits mistakes and profits from them. If you regard criticism as help and mistakes as part of experience, you will be able to figure out how to make improvements in your personal characteristics.

Cooperation: Getting Along with Others

Cooperation is the ability to work with others. Without it, modern society could not function. It is hard to think of an activ-

Cooperation is the ability to work with others in almost any situation.

ity that does not require cooperation. Even if you read or listen to music alone, people had to cooperate to publish the book or to make the record you enjoy. Composers, musicians, engineers, recorders, artists, writers, marketing people, store owners were all involved in the process of publishing the book or producing the record.

Try to think of all the people you cooperate with. There are your parents, neighbors, friends, classmates, teachers, employers, clerks in stores, and so on. In school you cooperate with your teachers. You want to learn so you can succeed in life. Their job is to help you learn. By doing your assignments and taking part in class discussions, you help the teachers do their job while helping yourself. Cooperation is a key characteristic for you as an individual, for the groups you belong to, and for the nation.

The opposite of cooperation is *conflict*. Conflict is not the same thing as *competition*. You might compete with your coworkers in trying to do a good job at selling so you would get a raise and promotion. But at the same time, you would want to cooperate with them to help make your employer more successful. If you started trying to undermine coworkers by talking behind their backs or refusing to cooperate, you would create conflict. This would hurt your employer's business and influence the status of your employment.

Most people seek jobs that require cooperation because they like to interact with others. Suppose you work in a fertilizer and seed store. You will have to cooperate with your supervisor by following directions and by making suggestions about how to improve store operations. With coworkers, you will try to avoid interfering with their work and to give them a hand when they need it. And you will cooperate with customers by helping them figure out which seeds are best for their type of soil, the kind of fertilizer that they should use, and when to plant their seeds.

Social Skills

Consideration for others is the basis of acceptable *social behavior*, the ways you conduct yourself with other people. Rules or guides for conduct have evolved over the centuries as people have worked and socialized in groups. Appropriate social behavior offers a common and easily learned way to feel comfortable and to help others feel comfortable. It involves good manners and also grooming and appearance.

Practicing Good Manners

A knowledge of *good manners*, proper ways of acting, ensures that you know what to do in social situations. This involves good manners in travel, in restaurants and other public places, and at the dining table and some general *courtesies*, or ways of being considerate to others.

Good Manners in Travel. You have been picked to attend the national leadership conference as a representative of your FFA chapter. You will fly there and spend a couple of nights in a motel. Good manners will make the trip pleasant for you and for those who serve you. Try to organize your trip. Make reservations early and arrive at the airport with plenty of time to spare.

Your planning will avoid inconveniencing other travelers or the people who serve you. Checking your luggage in early avoids the lines of last-minute travelers. You should board the plane quietly and quickly, without crowding others. Move quickly to your seat. Keep your seat area clean. If you must leave your seat, a simple "excuse me" shows your seatmates you care about their comfort and convenience.

Thank porters and others who serve you both verbally and with a modest tip. With your motel reservations made in advance, you can check in quickly and efficiently. Fill out the registration card completely. Notify the desk in advance when you are getting ready to leave so they can prepare your bill for you.

Good Manners in Restaurants. Good manners in restaurants are similar to those in travel. Find out if reservations are needed and make them. Wait quietly for the headwaiter or hostess to show you to your table. Generally, older persons are seated first. Make your selection from the menu and place your order quietly. Don't delay the ordering or change your mind after placing an order.

Express thanks for courtesies and services you receive. If you need help, ask for it quietly and politely. Being loudly critical is considered immature. If there is something wrong, the person serving you will be grateful for your whispered comments. A tip of 10 to 15 percent of the check is normally left under the edge of the plate for the person serving you. (Sometimes, when a charge card is used, the tip is added to the bill.)

Good Manners at the Table. Many people feel uncomfortable eating in restaurants and at banquets or dinners. The formality and the presence of people they don't know well make them uneasy. They feel they do not know how to act or which piece of table service to use first. Common sense is a good guide here. Table manners can be learned and practiced. Generally, it will be safe to follow the lead of the host or hostess in using tableware. You begin by using silver from the outside when faced with a formal setting.

Bad table manners can make those around you feel uncomfortable.

Proper eating is a matter of being quiet and inconspicuous and choosing small portions. If you are hungry after a first helping, you probably can have seconds. Keep your dishes neat and don't mix foods together on your plate. Bread is broken in half twice and buttered one piece at a time. Meat and other foods are cut into one or a few pieces at a time. After using the knife, place it on the edge of the plate.

Avoid messiness. Talk only when your mouth is clear of food. Remove bones and pits from the mouth with thumb and forefinger, using your napkin, and place them on the side of your plate. Do this as discreetly as possible. Here are some specific eating techniques:

- Soup: Dip the spoon toward the outer edge of the bowl, sip noiselessly from the side of the spoon. Don't blow on soup or drink it directly from the bowl.
- Salad: Eat with a fork. Lettuce is to be eaten; it may be cut with a knife.
- Bacon: Eat with the fingers if crisp; otherwise eat it with a fork.
- Corn on the cob: Hold it firmly in both hands.
- Spaghetti: Use a fork in the right hand to twirl against a spoon in left hand.
- Steak: Cut into small portions, a piece or two at a time.
- Olives, celery, fruits, pickles: Use fingers.

General Courtesies. You have traveled to your meeting and made it through dinner in good shape. Before the meeting starts, people are milling around aimlessly. You want to put people at ease. You introduce yourself to people who seem ill at ease. Greet them with a firm handshake and look at them directly when you tell them your name. Learn how to remember names. Repeat the name in your mind during and after the introduction. Make a practice of using the name in conversation: "It was good to meet you, Mr. Carver." "That's a good point, Lucas."

Generally, you stand up when someone is introduced to you. At this meeting, the state secretary of agriculture is the keynote speaker. Everyone stands when the secretary and other distinguished guests enter the room. Once the event starts, you listen courteously to the speak-

A proper table setting.

ers. You do not interrupt. Questions can be saved for the end of the speech. Don't criticize people in meetings. If they are making a mistake, take them aside after the meeting to offer your comments. Laughing at another person's mistakes is a sign of immaturity. The key is to consider other people's feelings as you want your own considered.

Grooming and Appearance

You want to make a good and lasting impression at meetings. Your appearance shows your opinion of yourself and of the people at the meeting. If you are dirty or sloppy, it may indicate you have a low opinion of yourself and of others.

Often, future employers attend meetings that you go to now. Agricultural leaders and business executives often attend important FFA meetings and 4-H events as guests or featured speakers. Your appearance will be the first thing people notice about you. They judge how you look as well as how you act.

Personal hygiene is important for your health. Others accept you more readily when you practice good hygiene. You need not be as fanatic as radio and TV commercials urge. But people who bathe and use deodorant daily are more pleasant to be around. Clean and well-groomed hair, clean nails and teeth, an erect and alert posture—these things convey an impression of health, vitality, and interest.

Clothes also affect the impression you make. Styles change, but keeping neat and clean is always in style. Blue jeans are accepted in many more social situations today than a few years ago. But a person who insists on wearing jeans to a formal occasion is thoughtless and inconsiderate. If you wear jeans to an interview, you may not get the job.

Use Your Judgment

Cleanliness, courtesy, and consideration are important to your relations with other people. You don't always need to follow the formal rules. You have to be sensitive to the situation and to the people around you. If the situation calls for formal courtesy, it is good to know how to behave. But if the event is more informal, you can get along by being considerate of the needs of others. Consideration for others is the key.

Good Citizenship

A *citizen* is a member of a country. Citizens have both rights and responsibilities. When you accept and perform your responsibilities, you are practicing *good citizenship*. In a democracy, nobody forces you to be a good citizen. It's up to you. But if democracy is to work, good citizens must assume their responsibilities. You have local, state, and national citizenship responsibilities. These begin at home, in school, or in your organizations. The FFA emphasizes good citizenship through its Building Our American Communities (BOAC) program and other activities.

Right now, good citizenship requires

that you spend your time in school productively. Your present and future responsibilities make it necessary to make the most of the opportunities that school offers you. At the simplest level, good citizenship means obeying the law and keeping out of trouble. But it goes beyond this. Obeying the law is a passive activity, usually involving a series of don'ts: Don't drive too fast. Don't steal. Don't go through red lights. Don't litter. Good citizenship also implies a positive, active approach. At its highest levels, it means assuming the responsibilities of leadership.

When you organize a service project at school, you exercise leadership and good citizenship. You might organize a project in history class to study the history of American agriculture. Taking the initiative to get the project started and following through on it are as important as what you learn from it.

Respecting Others

A citizen must respect the rights of others. The belief that one person's rights end when they interfere with the rights of others is the basis of democracy. Some people think these rights are absolute. But the right to free speech does not include the right to yell fire in a crowded building. The right to assemble does not permit you to hold a meeting on school property without permission. Nor does the right to free speech mean that people have to listen to the speaker. But the speaker must be given the opportunity to speak.

Good grooming helps you make a good impression.

Most rights are based on common sense and common courtesy. If you don't want to hear another speak, you can leave. If somebody intrudes on your rights, you can ask the person to stop. Nobody has the right to deliver a political speech in your living room without your permission.

You respect other's rights by not injuring them. Many laws try to protect people from injury. Traffic laws were written to protect people on the highways. Rules against running in the halls at school were also made to protect people from injury. Good citizens obey these laws whether they are being watched or not.

You also have a right to own and maintain property without losing it through theft or damage. The U.S. Constitution guarantees people that the government cannot take their property without "due process of law." Many laws are written to protect property rights. You expect people to respect the property you have worked hard to get—a car, a stereo, a camera, a hunting rifle, or a radio, for example.

Personal religious beliefs and practices are among the other basic rights that this country protects and respects. Many people came to this country to escape persecution for their religious beliefs. Persecution includes laughing at people's religious beliefs or ridiculing them.

Respecting National Symbols

The nation has rights too, and good citizens respect those rights. *National sym-*

bols are treated with respect because they were developed to represent or reaffirm the allegiance of citizens to the United States. You should learn why symbols like the flag are respected and how they are treated.

The U.S. flag is displayed higher than, and to the right of, other flags. It is raised at sunrise and lowered at sunset. When it is flown at half-mast, the flag is raised to the top and then lowered. You show respect for the flag by standing when it passes and by learning and reciting "The Pledge of Allegiance to the Flag." The flag is raised briskly and lowered slowly at ceremonial occasions.

The national anthem usually is played or sung during flag-raising ceremonies. You show your respect by standing when it is played and by knowing and singing the words.

National monuments represent important events, concepts, or persons in American history. They may be historic sites such as Plymouth Rock in Massachusetts or the Gateway Arch at St. Louis. They can be memorials such as the Washington Monument or Arlington National Cemetery.

Learn to respect these monuments by being careful not to do any damage or leave any litter when you visit them. Try to learn their significance. A trip to Jamestown, Virginia, the first permanent English colony in America, will be enriched if you study its history before the visit. How did the colonists practice agriculture? What crops did they grow? What tools did they use?

Respecting the Environment

The Meadowbrook FFA Chapter decided to make the environment one of the themes for its yearly activities. One member reported on the environment as part of the national heritage of the United States. Another reported on the importance of protecting the quality of the air, land, and water.

The chapter organized a cleanup campaign and established a recycling center where papers, bottles, and cans could be brought. As part of the project, members checked on wasted energy use around their homes. They inspected and repaired inefficient furnaces and appliances. They checked their homes for insulation. The members came to realize that the environment included everything around them, not just the wilderness.

Maybe you can organize some environmental projects for your FFA chapter. The BOAC program can be used to sponsor such projects as cleaning up streams or establishing a wildlife refuge. The good citizen works to protect the environment by not degrading it in the first place and then by pitching in to clean up areas that are dirty.

Making a Contribution

Good citizens do not expect a free ride. They try to make a contribution that will improve their nation, community, organization, or family life. The first step is to be informed about the issues that affect your FFA chapter, your local area, your state, and the nation. What are the pros and cons of making a local stream a wildlife area? Where does the candidate for county commissioner stand on taxes on farmland?

You learn about political issues from newspapers, magazines, and news programs on radio and TV. Political parties provide you with information on their positions. Nonpartisan organizations like the League of Women Voters make detailed studies of issues. Interest groups provide information on the environment, consumer affairs, agricultural policy, or most other issues you want to learn about.

You need to evaluate the information you receive. Who is speaking? What point of view does the speaker represent? Is the information reliable, or do you need to check further?

As you study and learn the issues, you can work on political campaigns. In addition, you may want to communicate your views to elected officials.

The FFA's BOAC program focuses on community service. You can also work for nonschool organizations or with groups that have youth auxiliaries such as Circle K and the Key Club. You will multiply your contribution by working with others. Laws are not made and issues are not settled by individual effort alone.

As a good citizen, you will abide by the laws and decisions that were made by the majority, just as you do in your FFA chapter. Democracy will not work otherwise. If you help make a decision, you will be more willing to abide by the outcome, even though you might not agree with it 100 percent. If you do not agree with a decision, you can work to change it. But it deserves your support until it has been changed.

Working in the FFA helps develop the characteristics you can use to become a responsible member and leader. The responsibilities include attending all meetings, forming views, voicing opinions, as-

A youth leader showing good citizenship by giving blind and handicapped city children a chance to ride and care for horses.

suming and completing tasks, and abiding by the will of the majority.

Getting It All Together: A Review

Personal development is a lifelong process. A Chinese proverb says: "A journey of a thousand miles starts with a single step." Life is such a journey. If you have not already done so, you can take your first step on the journey of personal development right now. It may be the most important decision you ever make.

Personal development can be analyzed in terms of desirable personal characteristics, accepted social behavior, and good citizenship. You can choose to be as you want to be in each of these three areas. You shape your own development. No one can force you to be courteous or to act as a good citizen. You must decide and then do it. If you fail to plan your development, you shape it poorly. You can develop a list of characteristics you want to improve. Then work hard to improve them.

THINKING IT THROUGH

1. What personal characteristics are important for working and for getting along with people?
2. Describe why each characteristic identified is important.
3. Why are good manners important?
4. What manners should be practiced in traveling?
5. What manners should be practiced in restaurants?
6. What manners should be practiced at the table?
7. What general courtesies should be practiced in dealing with others?
8. Why is good grooming important?
9. What grooming practices should be followed?
10. What is good citizenship?
11. Identify ways to develop and practice good citizenship.
12. How does your FFA chapter help you to become a good citizen?

CHAPTER 4

MAKING DECISIONS

The school year is ending. You have offers to work for the summer on a dairy farm or in a farm supply store. What should you do? You are faced with a *decision.* This simply means that you must make up your mind and choose between different courses of action. If you fail to reach a decision, you will get neither job and may spend the summer with no work and no income.

The ability to make decisions is the mark of an effective person. It also is the quality of an effective leader. People who know how to make decisions tend to get along well in life. People who cannot make decisions don't go very far. But you make decisions all the time, whether you think about it or not. What shall you eat? What should you wear? What will you do today, tomorrow, next week, or the rest of your life? You have to live with the results of the failure to make a decision as much as with a bad decision. However, the goal is not just to make any decision, but to make the best possible decision.

Take that summer job, for example. The job at the dairy farm allows you to work outdoors and with animals. The farm supply store would give you a chance to work with people and learn about the business. Both jobs could lead to work after graduation. Which one suits you best? To make a good decision, you have to know how to weigh *alternatives,* the various options you have available to you. The process of making decisions is critical in life and leadership.

CHAPTER GOALS

In this chapter your goals are:

- To evaluate why it is important to make sound decisions
- To analyze how the process of change can influence decision making
- To outline a plan to follow in making decisions

Decisions Affect Your Life

Making decisions is one way to take responsibility. In making a decision, you say what you want and what you are willing to do to get it. You may have faced some difficult decisions already. You have had to decide which courses to take and which organizations to work with. Should you take vocational agriculture or industrial education? Do you want to put your energy into 4-H, the FFA, or the Boy Scouts of America? Should you try out for the volleyball or basketball team?

Were these decisions hard or easy?

Leaders are judged by their ability to make decisions.

People react to making decisions in different ways. For some people, every decision is an agonizing experience. They are afraid to choose and perhaps are not clear about the alternatives. For others, making a decision is relatively easy. They have a clear grasp of what they want and how to get it.

You already have grappled with some important decisions. Before choosing your courses, you had to decide which careers interested you. You also decide how you are going to earn money and how to divide up your time. You can get some idea of how well you make decisions by looking at how you decided those issues. Before long, you will face other types of decisions. What job will you take? Will you go to work immediately or seek more education? Where will you live? When will you marry? These are long-term decisions that influence your life.

It takes maturity to consider all of the factors involved and then to make wise decisions. Some people make decisions without thinking things through. Others have a hard time making up their minds. The mature person looks at all the factors and then is able to make an informed decision that is likely to bring about the desired result.

You must be able to make decisions for yourself before you can make decisions involving others. And *everyone* makes decisions, not just presidents and business executives. Decisions affect your life every day. It is important to learn to make good ones—first for yourself, and later to help others as a leader.

Decisions and Leadership

Leaders often are judged by their ability to make decisions. Try to think of presidents of your class or FFA chapter who did a good job. You will probably find that they were able to make decisions. Maybe they weren't always right, but they dared to make decisions. There's a good chance that anybody you admire is *decisive,* is able to direct his or her own life effectively.

To make good decisions, you must start with a sense of your own system of *values,* the things you feel are important. How will the decision relate to your values? What do you want to accomplish? In choosing a summer or part-time job, is it more important to make a lot of money or to learn about a career? As leader of an FFA chapter, would it be more important to get more members involved immediately or to work to build an exciting program?

A second element of decision making is a knowledge and understanding of goals and needs. In seeking work, you should be aware of your own needs for income or self-satisfaction. These same needs might be linked to a goal of getting additional education after high school. In a group such as FFA, both leaders and members must know and understand the goals and needs. What does the organization want as a collection of individuals? What are its needs as a group? Perhaps it needs to raise money to carry out projects.

Sometimes the goals of an organization are written down in its charter or bylaws. Usually, the goals and needs of the group must be determined. One of the marks of leadership is finding out what the members want and helping them decide how to get it.

In carrying out the goals and wishes of the group, the FFA or Scout leader must know how to deal with two kinds of decisions. One kind will be made by the leader for the group. The other kind will be made by the group as a whole.

Leader Decisions. The membership of a group cannot be expected to vote on every decision that affects the group. One of the reasons there are leaders is to make day-to-day decisions. For example, an FFA chapter would not vote every time a news release is to be mailed to the local newspaper and radio station. The leaders could make that decision. Meeting time would be saved for larger decisions such as the planning for the year's FFA banquet.

Often, leaders choose members to serve on committees or to do other jobs within the organization. They also frequently select the time, place, and date for meetings and decide which items will be discussed and in what order. For example, if you were chairperson of the committee to plan recreation after the next FFA meeting, you would decide when and where the committee meeting would take place and notify the members.

Group Decisions. In the United States, most groups operate democratically. Important decisions are made by members of the group. Very large groups may choose representatives to make group decisions. Perhaps your school has an elected student council that makes decisions for all students.

The leader's job is to help the group make a good decision. One important function of a leader is knowing which decisions should be made by the group and which by the leader. The leader can decide when a committee is to meet, but the committee would decide which projects to recommend to the membership.

The leader can help the group make decisions by clarifying the issues, presenting the facts on which decisions are based, and making sure that there is an orderly process for making the decision. To have an active and vital group, it is important to promote group decisions.

If the leader makes sure that all points of view are heard, hidden disagreements can be brought into the open before the decision is made. If disagreements and opposition remain hidden, the leader will have a hard time carrying out the decision.

Decisions that are made by a group process are more likely to be supported by all members of the group. Those who still disagree with the decision will at least feel that their ideas were heard and considered. Those who support the decision will do so more enthusiastically because they helped to make it. Many businesses believe that people will work harder to make decisions work if they have helped make the decisions.

Following Through on Decisions. The process of making a decision places a responsibility for carrying it out on those who made it. While the leader may be delegated with power to carry out the decision, the leader cannot do it alone. Help usually is needed. Leaders should work to help the group accept responsibility for decisions. For example, if your FFA chapter takes orders and buys fruit to sell at a fair, the fruit must be delivered promptly or the chapter will lose money. Taking responsibility for the decision to buy and sell fruit means following through on the decision and making sure the fruit gets sold promptly.

You must also take responsibility for your individual decisions, both for carrying them out and for their consequences. If you decide to become a soil scientist, you must study the proper materials and take the proper courses to follow through on the decision.

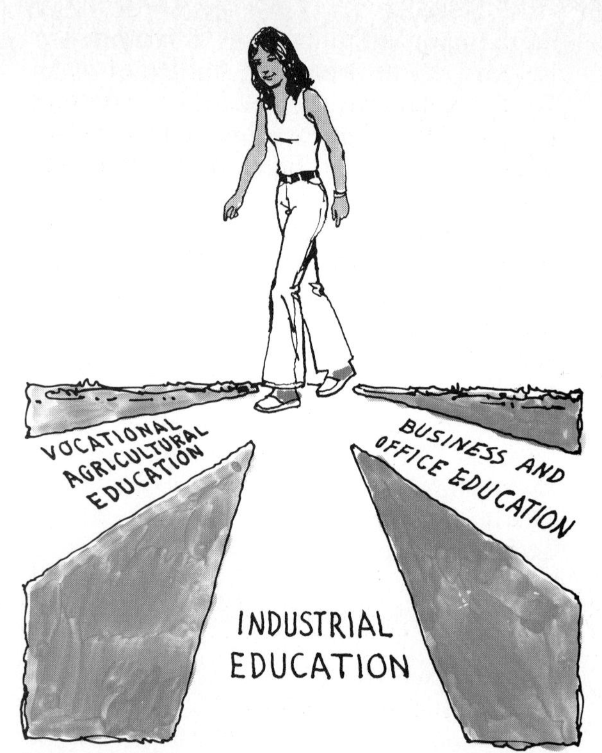

Making a decision is one way to take responsibility.

Decisions and Change

Most decisions lead to action or to a project. For example, your FFA chapter might plan a field trip, a fund-raising event, or a beautification program for your community. However, groups often face tougher kinds of decisions. These are decisions to act to meet changing conditions. In the past few years, many formerly all-male groups have accepted females into membership. In some cases, this change required many months of planning and effort. The FFA and the Boy Scouts of America recently made such a change.

Making changes of this kind can be frightening because nobody knows what the group will be like afterwards. People tend to hold on to the familiar. This has been shown when organizations have tried to change and accept women and minorities as members, associates, and employees. For many years, the FFA failed to accept females as members. It was feared they might dominate and distract from organizational goals. However, many have proved to be valuable members of the organization.

Similar fears are raised by the admission of women to careers and positions of leadership that formerly were closed to them. Men who hold these jobs or positions fear they will be pushed out. When difficult decisions about change are made, issues such as this must be dealt with openly.

If leaders are to help their groups change, they must understand the change process. Change is inevitable in a dynamic society. Alvin Toffler argued in a book called *Future Shock* that there has been so much change in our modern, dynamic society that people cannot deal with it; they suffer from "future shock." Good leaders can help the people they are leading go through necessary changes without too much shock or suffering.

How Change Occurs

Social scientists who have studied change have described the change process as including three phases. First, an effort must be made to "unfreeze" the present level and to develop a willingness or a need to change. This unfreezing can be accomplished in a number of ways. Minorities helped unfreeze the present by an appeal to the nation's democratic ideals. They argued that it was not fair or democratic for members of their groups to go to certain schools, live in substandard houses, or earn less money for the same amount and kind of work. Sometimes, the present can be unfrozen by new information.

Information also is the key to the second phase: moving up to a new level. The leader for change supplies the group with information or ideas that are needed to bring about the change or to make it possible. In agriculture a century ago, the land-grant colleges and county agents used new information to help farmers move to a new level. Farming had not changed much for thousands of years. The land-grant colleges developed new methods of farming: improved machinery, new fertilizers, new seed strains, different ways of tilling the soil. The county agents shared this new information with farmers. The agents often proved the superiority of a new technique by having experimental plots planted and by asking key farmers to try the practice. The system revolutionized American agriculture.

The third phase is "freezing" at a new level. The group recognizes the new situation. The group members and the leaders commit themselves to operate under the new set of circumstances. Leaders must recognize the importance of their role in the change process, because some members will not agree with the change.

Conditions Influencing Change

Almost as much has been written about change as about leadership. One thing that has been learned is that if a change is to be accepted, those to be affected by the change should help plan and decide to make it. For example, workers are more likely to accept new labor saving machinery in their plant if they have helped make the decision to introduce the machinery. They will know why the change is needed and what it will mean to them. Will it increase company profits and perhaps their wages? Will it make their jobs easier? Will anyone be displaced by the machinery? What new skills will have to be learned?

To pave the way for accepting change, leaders must work for an open climate. This can be done best in the context of a group decision, where everyone can hear and be heard. Detailed factual information must be provided to members to support the need for change. Why should girls be permitted to join your FFA chapter? How will that affect you? What are the benefits of this step? The drawbacks? The disadvantages of change should be discussed as honestly as the advantages.

To effectively help a group change, a leader needs to know about the following areas:

The Leader's Status. Is the leader liked and respected by the group? Does the leader have to earn the group's trust? Is the leader seen as a strong, decisive person who can make decisions or as someone who does only what the group or a faction of the group wants?

If the leader is seen as a strong, independent person, the group will tend to trust her or him. If the leader is the captive of one faction, then the rest of the group will not trust his or her leadership.

What the Group Knows. A group may have different kinds of information. Some may be accurate and factual, but there may not be enough of it. Other information may be inaccurate and misleading. The leader needs to know how much of both kinds of information the group possesses.

The FFA chapter was deciding whether to buy a popcorn popper to raise money. Several members opposed the idea. They argued that it had not worked in other schools. When the committee reported, it was found that the information of the members had been inaccurate. The committee presented the cost, a payment plan, and letters from other chapters stating that money had been raised with popcorn poppers. With the new information, the chapter voted yes.

Presenting New Information. Conditions change. It is important to be flexible in making decisions. Good leaders encourage groups to change in the light of new information. In presenting this information, the leader should be aware that

The process of change as shown in the mechanization of cotton picking.

people differ in the ways they learn best. Some learn best from the spoken word, others through reading. An all-male group that is considering a change in its bylaws to admit females will need plenty of information. Leaders should present it both orally and in writing so everyone will know exactly what has been, and must be, done.

Informal Communications. All groups have informal ways in which members communicate with each other. Members discuss with others what is on their minds. The leader needs to be aware of this network and how it works. Which members do others talk to most? Which ones are most influential? Which ones tend to be sources of new information? The leader will want to know who the opinion leaders are and to be sure that correct information is available to them.

Promoting Group Interaction. In dealing with change, group members should be encouraged to express their thoughts and feelings about the proposed change with other members and with leaders. This process will help the group become more comfortable with the proposed change and may bring problems to the surface before, rather than after, the decision. Perhaps you are presiding at an FFA meeting, and nobody is speaking up on an important issue. You know some members have strong feelings about the subject. You could appoint them to a committee and ask them to report back. Or you could break the meeting up into small groups so that everybody could have their say. A recorder would be appointed for each group, which would report back to the whole meeting.

Making the Change

If all conditions for promoting change have been dealt with adequately, then a decision can be made. The time after the decision is as important as the decision itself. This is when the decision to change is carried out. Making the change must be done carefully and in the context of the group decision. Part of the decision making should deal with what happens once it is made. How will the change be made?

The wishes of the group must be followed. At this point, the leader is as important as at any other time. This is what it means to be an *executive*. You execute, or carry out, the decision of the group. Your FFA chapter may decide to change the local dues. Much of the work of carrying out that decision will be left to the treasurer who collects the dues.

Making Good Decisions

President John F. Kennedy was fond of saying that "things don't just happen; they are made to happen." Things are made to happen by good decisions. And just as change can be studied, so can the way decisions are made. You can learn how to handle change and how to make decisions. If you can make decisions, you will probably be happier as an individual, and you may become a leader and guide groups in making decisions.

It takes practice to learn to make good decisions, much as it takes practice to learn tennis, English grammar, or welding. Success in making good decisions leads to confidence in your decision-making ability. Therefore, you should be aware of the decisions you make and take care that they are made properly.

Steps in Decision Making

It may be easier to understand how decisions are made if you break the process down into the following steps:

This FFA chapter has to decide between two alternatives for their banquet.

1. Define the Problem. A problem does not necessarily imply something unpleasant. It could be a choice between two good jobs. To define a problem, you must focus on the specific points that are needed for a decision. What are you trying to decide? What problems are you trying to solve, or what goal are you trying to reach? Maybe your FFA chapter is planning a banquet. The problem is to choose whether to have the meal prepared at school or catered from a local restaurant.

2. Identify the Facts. It is helpful to outline all the facts. The facts in our FFA problem are that the banquet will cost $1 more per plate if it is catered and there is only enough money in the treasury to cover the cost of a school meal.

3. Plan Possible Solutions. When you face a problem or a number of choices, there are usually a number of possible decisions. To make good decisions, you need to look at as many alternative solutions as you can. Your FFA chapter could have the school prepare the meal, secure funds from members for the added cost of a catered meal, or try to raise money with a project.

4. Evaluate the Solutions. This means you must look for the best possible answer. The evaluation includes looking at the advantages and disadvantages of every solution. Your FFA chapter would discuss where to have the banquet. The main points could be written on a chalkboard. There is not enough time to raise money. Most members seem to feel that making them pay extra would be unfair.

5. Make the Decision. After going through the first four steps and weighing all the advantages and disadvantages of all possible solutions, you make your decision. You say that on the basis of all these considerations you choose this course of action over all others. Choosing always involves taking responsibility. Failure to choose leaves your life in the hands of other people.

If you don't make a decision on the FFA banquet, then it will not be held. If you choose the restaurant and do not provide for the extra funds needed, then your chapter will be in trouble. In this instance, based on the steps of the process, your chapter votes to hold the banquet at the school.

These five steps can work for a group or for an individual. In a group situation, the leaders must be aware of the steps and guide the group through each of them. If the person with the title of leader doesn't do this, the group may flounder and make a bad decision—or none at all. If somebody else understands the process and helps the group focus on the important issues, then that person is the leader, no matter who has the formal title.

Importance of Decisions: A Review

Individuals and groups make decisions all the time. A person makes many kinds of decisions. The decisions you make now will influence what happens to you in the future. Maybe you think keeping records is unimportant to your goals. But if you later decide to apply for a state or national award in the FFA, you will need a good set of records.

Decisions also affect groups in different ways. Some affect the way the leader performs. Some affect the group and its wishes and goals. Some improve the way the group functions. By taking the time and effort to analyze each problem, individuals and groups will vastly improve the chances of making good decisions.

THINKING IT THROUGH

1. Why is it important to be able to make sound decisions?
2. Describe how decision making relates to leadership.
3. Why are group decisions important?
4. How does change occur?
5. What knowledge does a leader need to promote change?
6. What steps should be followed in preparing for and making a decision?

UNIT II

WORKING WITH OTHERS

COMPETENCIES

Competencies	PRODUCTION AGRICULTURE Livestock farmer/rancher	AGRICULTURAL SUPPLIES/SERVICES Grain elevator manager	AGRICULTURAL MECHANICS Machinery service manager	AGRICULTURAL PRODUCTS, PROCESSING, AND MARKETING Grain buyer	HORTICULTURE Nursery worker	FORESTRY Timber faller	RENEWABLE NATURAL RESOURCES Park worker
Demonstrate good human relations	Very Important	Very Important	Very Important	Very Important	Very Important	Important	Very Important
Show an ability to get along with others	Very Important	Very Important	Very Important	Very Important	Very Important	Very Important	Very Important
Exhibit the ability to work as a member of a team	Important	Very Important	Very Important	Very Important	Very Important	Very Important	Very Important
Communicate effectively using the phones and mail	Very Important	Very Important	Very Important	Very Important	Very Important	Important	Very Important
Develop group goals	Very Important	Very Important	Very Important	Very Important	Important	Important	Important
Consult with your supervisor, coworkers, and others	Very Important	Very Important	Very Important	Very Important	Very Important	Very Important	Very Important
Demonstrate the ability to participate in group meeting	Very Important	Very Important	Very Important	Very Important	Very Important	Very Important	Very Important
Make appointments with others	Very Important	Very Important	Very Important	Important	Very Important	Important	Very Important
Develop abilities to communicate effectively with individuals and groups	Very Important	Very Important	Very Important	Very Important	Very Important	Very Important	Very Important

Very Important

Important

Not Important

COMPETENCIES	PRODUCTION AGRICULTURE Livestock farmer/rancher	AGRICULTURAL SUPPLIES/SERVICES Grain elevator manager	AGRICULTURAL MECHANICS Machinery service manager	AGRICULTURAL PRODUCTS, PROCESSING, AND MARKETING Grain buyer	HORTICULTURE Nursery worker	FORESTRY Timber faller	RENEWABLE NATURAL RESOURCES Park worker
Demonstrate the ability to cooperate with coworkers							
Demonstrate the ability to use parliamentary procedure when participating in a meeting							

Most of the important parts of your life will be spent in group situations. It is hard to think of a situation which does not lead eventually to interaction with a group. Even when you read or study alone, you usually are aiming to improve your relationship with some group. Your studies are shared with your classmates, and your learning helps you work with groups as an employee or as a responsible citizen.

You started life in your family group. Soon your world expanded to include the group of neighborhood children. As you went to nursery school or kindergarten, you experienced your first structured group situation. Later comes grade and high school, perhaps a vocational technical school or college, work, and affiliation with a variety of organizations. All these experiences prepare you to work and to live a productive life. The matrix on the facing page shows you some important abilities you need in working with others as they apply to specific jobs.

Stop and think. How many groups do you belong to now? How many have you belonged to in the past? Don't forget the informal groups, such as the regular touch football group, the group that attends movies together, or the one that gets together after school to study or just chat.

Since you spend so much time in groups, it pays to work to increase your effectiveness in group situations. This is vital if you are to have a full and satisfying life. As you move from group to group, you will act as both a leader and a member. Maybe you won't be a formal leader, but you certainly will act as an informal leader.

So you will want to develop the skills needed for effective participation in groups. Effective participation requires a complete understanding of communication on both individual and group levels. Communication is the basis of human relations and group interaction. Groups cannot function unless members have learned basic communication skills and use them effectively. To participate in a meeting requires some skill in communicating with groups. Presiding at a meeting and public speaking involve more difficult communication skills.

Taking part in group activities involves you in what is called the group process or group dynamics. This means that groups act and react in certain ways, just as individuals do. There are skills and techniques that you can learn that will make the group process work better. You will find that there are many roles you can adopt as a member and leader of the group that will help it function more effectively.

One of the important ways to make groups work better is to understand and use parliamentary procedure. Parliamentary procedure is designed to ensure that groups function democratically. With proper parliamentary procedure, everybody gets a chance to be heard, and the rights of the minority are protected. At the same time, the majority is able to make a decision.

Your overall goal in this unit is to learn how to communicate and to participate effectively in a variety of group settings. This will serve you now and prepare you for work and community participation.

CHAPTER 5

HUMAN RELATIONS

Sam was new at the feed mill and wanted to do well. When his work as forklift operator was slow, he would find other jobs to do or ask questions. His interest pleased the mill owner, who knew the value of initiative and enthusiasm in his employees. He often gave Sam a pat on the shoulder or some praise. One day, Willis, another employee, said to Sam: "You'd better watch yourself. Some people don't like how you're trying to get in good with the boss." This warning worried Sam. Willis had worked at the mill for years. He did odd jobs and seemed liked by other employees. Sam was not sure what to do. He wanted to learn more and be responsible. But he wanted the friendship of his coworkers.

Sam had a very important human relations problem. To keep a job, you need to get along with your supervisor and coworkers. *Human relations* deal with how you get along with others. You are often in contact with people. You have human relations with family members, friends, classmates, teachers, coworkers, supervisors, and many other people.

The quality of your human relations determines the quality of your life. Human relations are important for very practical reasons. Most people who are fired are fired because of their inability to get along with others. No matter how good you are as a mechanic, if you can't get along with coworkers you won't keep your job. Human relations is a two-way street, but you only control your end of the street. How you feel and act affects how others feel and act toward you.

> Sam asked Harriet, a milling-machine operator, about Willis's warning. She said: "Don't let Willis bother you. He told me that workers resented me because I was the first woman hired here. Willis's lack of initiative has kept him in the same job for years. He gets jealous when he sees someone else trying to get ahead. If you do good work, you will be respected by the supervisor and the other workers."

Human relations rules may or may not be spelled out. You probably know exactly when to be quiet and when to speak in your classroom. This may not be as clear on the job. However, your employer may have certain requirements for dress and grooming, while your school is more informal.

Often, you will have to figure out what is expected from what the other person is like, from the situation, or from past experience. In doing this, you may find you can't get along with everyone. But if you have trouble with most people, take a look at your own attitudes and behavior. Something may be lacking in your approach to human relations. It's easy to blame others. But only *you* can change the quality of your own human relations.

Everyone has some good human relations skills, but may lack others. You may be punctual and courteous, but find it hard to meet people. You are responsible for honestly evaluating how you get along with others and then seeking ways to improve your human relations skills.

CHAPTER GOALS

In this chapter your goals are:

- To describe the importance of working with others as a team
- To develop the ability to get along with others as individuals
- To analyze the keys to understanding yourself and others
- To develop ways to deal with prejudice

Developing Human Relations as a Leader

To develop good human relations, you need to work toward three basic goals that are also essential to leadership. These goals are to promote cooperation, to create a desire to work, and to promote satisfaction with a situation.

Promoting Cooperation

Cooperation is working together for a common goal. It is a key to success in a school, a business, or an FFA chapter.

When there were jobs to be done, Trish would figure out the person best suited for each task. As head of the FFA field-trip committee, Trish suggested that Robert call places they wanted to visit because he liked to deal with people. She knew Lewis was working on his public speaking, so she suggested that he make the oral report to the chapter. Trish was able to get people to work together because she was sensitive to their needs and abilities.

Creating a Desire to Work

Group members must be motivated if they are to work together for the good of the group. Skill in human relations can help create a desire or motivation to work through a variety of rewards or incentives. People may be motivated to work by the desire to do a good job, by money, by the need for friendship, or by the satisfaction of being in charge. Which of these factors motivates you?

Mr. Nogales knew that a mixture of incentives produced the best results. His employees at the Brown Swiss Dairy Store liked to work hard for him. Keith loved to find out what was wrong with machines and fix them. Mr. Nogales made him maintenance manager. Rowena, who enjoyed supervising and planning, was named supervisor of the production staff. Toby was assigned to the cash register because he liked people. Mr. Nogales also knew the value of praise: "Another fine batch of cottage cheese; you make the best in town." And he used other incentives. Hard work over a period of time could bring a bonus or a day off.

Promoting Satisfaction

Satisfied people work harder, learn more, and get along better with others. They make good employees.

As ranch manager, Ted spread the hard and dirty jobs around. If somebody was told to clean the barn, he would say: "I know this isn't the most enjoyable job, but it has to be done. Do it well today, and I'll ask somebody else to do it next time."

Promoting cooperation and satisfaction and creating a desire to work are important in all situations. The results may be harder to measure in less formal settings. But if you feel good about the situation, about yourself, and about others, your human relations are probably in order. If you feel angry or discontented, ask yourself if you are doing anything to create the unpleasantness.

Acting Positively

While fixed rules do not apply to every situation every time, the following suggestions may help you learn how to get along better with others:

Human relations are important in almost every aspect of your life.

Inspiring a team to work together is using human relations to get cooperation.

Think before you speak. Try to understand a situation before speaking out. This is a responsible posture. And don't criticize needlessly. Everyone makes mistakes. If it's necessary to criticize, do it tactfully, carefully, and privately. Making fun of others is always bad human relations.

Praise a job well done. This almost always works—even with those who say they don't like praise.

Be interested in others. Ask others what they are doing. What are they interested in? People enjoy talking about themselves. This human relations practice also can uncover hidden talents. As an FFA leader, you may make members active by showing an interest in each one.

Be cheerful and pleasant. Speak to others. Silence may be interpreted as dislike. Little is lost by saying "Hello, how are you?" as an experiment. You may find a shy person who would like to have a friend.

Let your strong points speak for themselves. Don't brag. Say what you can do and then do it. If you do a good job with an FFA task, you will be given more responsibility. The same is true on the job. "Show, don't tell" is a good rule.

Consider others' feelings. If a normally cheerful person is in a grumpy mood, there probably is a good reason for that mood. You may want to ask what is wrong, but don't push it. Respect a desire to be left alone.

Keep your promises. Much business is done on the basis of a word and a handshake. If you don't keep your word, people won't trust you. Make few promises, but try to do each job as it comes along, and then look forward to the next one.

Don't demand praise. If you do good work, praise will come. Be satisfied that you are doing good work and that your work is noticed. Some supervisors believe that too much praise is not good.

Don't take it personally. Ignore those occasional ill-natured remarks. Try to understand the other person. Often, the comments reflect a problem of the person making them. This is especially true if the other person is generally cheerful and friendly. If such comments become unbearable, tell the other person how you feel. Or talk it over with a teacher, supervisor, or friend. Seek a solution.

What Makes People Tick?

Certain characteristics and suggestions may help your human relations. But there is no substitute for understanding people. What do they need and want? Why do they act the way they do? What do they want from you? If you can answer these questions, you will know how to improve your human relations.

Understanding Yourself

First, you need to know yourself. You are made up of different motives (reasons for acting a certain way), thoughts, feelings, and reactions. So are other people. In 1954, the noted psychologist Dr. Abraham Maslow developed the theory called Maslow's "Hierarchy of Needs." His theory describes five human needs that are basic to all individuals. It says that all actions are a result of these basic needs. These needs are given in the order in which they must be satisfied.

Survival. Artis constantly skipped meals and got too little sleep. He skimped on his need to survive—to stay alive. He was often irritable and found it hard to pay attention. By shortchanging his survival needs, he hurt his human relations.

Keeping healthy is part of keeping alive and must be done if the other needs are to be met. According to Maslow, the need to survive is a person's primary concern.

Security. Security is freedom from fear. A person will try to satisfy this need once survival needs are met. Judy took care of this need. She knew that a good job and living situation were important for her security. Her sales job at the farm supply store gave her security, and living with friends provided her with companionship and protection. She sought freedom from fear.

Love and Approval. Alice felt best when she was with family or close friends. Her love was a commitment and a desire to please those she was close to. She shared happy and sad moments with them. When one of her friends had a hard time, Alice stood by loyally and tried to help. She took part in many FFA activities because this brought pleasure and approval from her friends. Love and approval, then, is the next level of need according to Maslow's "Hierarchy of Needs."

Respect. Thomas worked hard in his vocational agriculture classes. He wanted the respect of the teachers and the admiration of his classmates. He also took part in many school activities. He worked hard in FFA, the music club, and the basketball team. Tom earned the respect of others, thus meeting Maslow's next basic need.

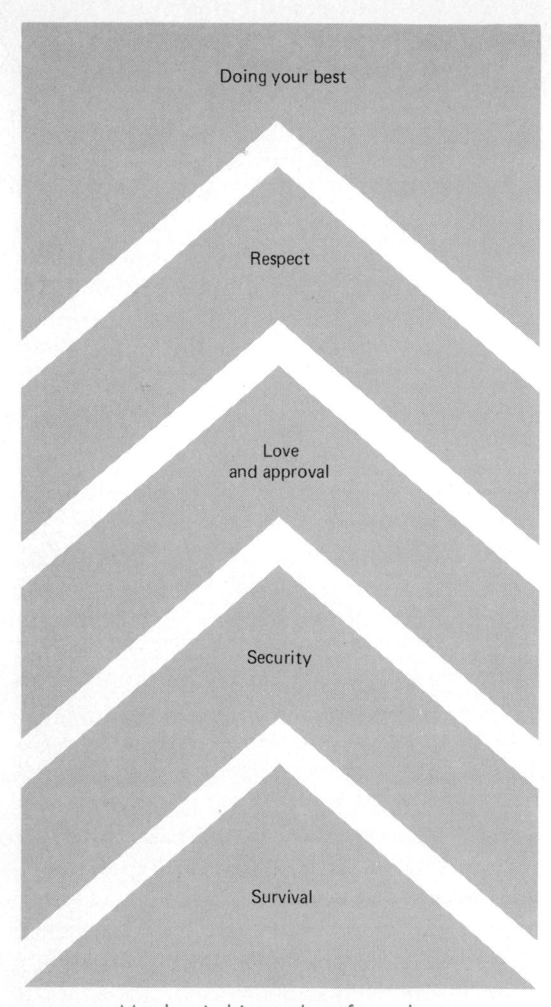

Maslow's hierarchy of needs.

Self-Realization. Self-realization means that a person's potential has been realized—that a person has done his or her best. Marcus never did his best because he didn't know what his best was. He had no idea of his potential. He thought he had no skills. But he loved to work with tools. He often fixed small engines on his parents' farm. His teacher said: "Mark, you have a great potential for mechanics. Why don't you take a couple of agricultural mechanics courses? Mark took the teacher's advice. He did have a potential for mechanics. His new interest led to placement in a tractor dealership. He found satisfaction in working hard at a job that he could do well and that he liked.

Doing your best is the highest need level. People work at satisfying this need after their others have been met.

Understanding Others

Other people have the same needs as you. The strength and flavor may be different, but the basic components are the same. Understanding this is important to human relations. But you won't always know which need a person is meeting.

> Jack skipped lunch. It made him irritable, but he wanted to fill a large hardware order before it was picked up at 2 p.m. Charles, his supervisor, urged Jack to take a lunch break and was surprised when Jack got angry. Jack was postponing his basic need to eat to fulfill his need to do his job to the best of his ability.

Benefit of the Doubt. It helps to give others the benefit of the doubt. It may be tempting to judge someone who is different as strange or bad. Everybody makes excuses for themselves. But what about the other person?

> Alicia had a headache and was permitted to go home early, even though her work was not finished. Next day, Alicia found that some inventories had not been completed in the stockroom. She said: "That Pedro, he never does his work. I wish he would work harder." Alicia had given herself the benefit of the doubt, but not Pedro. Pedro's child was sick and he had spent the day taking care of him.

It is important to know the facts. People have reasons for acting the way they do. The reasons may not always justify the actions, but knowing there are reasons will help you accept other people.

Consider the Whole Person. Laura and Doug worked together on their FFA chapter's program committee. Their cheerfulness made other people want to contribute to the committee. Then Doug got grumpy. People stopped coming to committee meetings to give their ideas. This made Laura very angry. She mentioned her anger to Nathan. He said Doug's father was sick and that Doug had a lot of responsibility at home. He felt a lot of pressure to finish his schoolwork, his chores at home, and his FFA tasks.

This knowledge enabled Laura to see Doug in a different way. He was not just a friend and committee member. He also was a son, a family member, and a student. With this understanding, Laura suggested that Doug ask to be relieved of his committee responsibilities until his father got well.

You may often forget to consider all aspects of another person. But instead of assuming that a person who is being difficult is bad, you might consider the possibility that something is going on that you don't know about. Another part of the person's life is affecting the part that touches your life.

Other People's Needs. Other people have needs too. When you are tired, you go to bed. But sometimes people are not aware of the needs on which they are acting. You can use this knowledge to improve your human relations.

> Seth and Walker covered agriculture for a newspaper. During slow periods, they would chat. Walker bragged a lot. He had the best car, the most beautiful home, the brightest children, and so on. Seth felt that any topic of conversation always led to a chance for Walker to brag. Then Seth realized that Walker was insecure and needed approval. He began to compliment Walker on his good newspaper stories. Over time, Walker began to feel that Seth respected him. He did not need to brag so much.

Dealing with Prejudice

Knowing that all people have the same needs and characteristics helps us do away with *stereotypes,* which are blanket judgments about classes of people. How often have you heard statements like: "Girls are sissies; they can't play ball"?

Statements like that are based on *prejudice,* which means prejudging people before you know them. Prejudice paints everyone in a group with the same brush. Look out for statements that begin: "All city people. . . ." Or "All older people. . . ." Or "All country people. . . ."

Prejudice is a way to put people down. It makes them feel small. Since everyone shares characteristics with other people, everyone will probably be put down along with the groups they belong to. When it happens to you, you can see that prejudiced statements don't hold water. When were you put down? Did somebody react to you with prejudice? "Country people are slow and stupid." "You know how awkward left-handers are."

Wisdom Starts at Home. The beginning of wisdom in matters of prejudice is within yourself. How do you feel about fat people, for example? They may or may not be jolly. If you accept the prejudiced idea that blondes have more fun, you may limit your capacity for enjoyment if you are not a blonde. What stereotypes do you accept

This 4-H member understands the needs of the younger member and is sharing a project with him.

as true? Are there classes of people you don't like? Do you belong to a group that is the target of prejudice?

The tragedy of stereotypes is that people tend not to look beyond them. You may well find fat people who are jolly, if you have that prejudice. But blinded by your prejudgment, you may not notice that the person also is generous, kind, poetic, competent, sad, intelligent, mean, witty, talented, manipulative, and sometimes angry. In fact, that person is a mixture of characteristics that make up a far more interesting human being than can be described by a stereotype.

Prejudice and Discrimination

Prejudices often serve as the basis for *discrimination,* or unfair treatment. The stereotype of "women's work" narrowly limits the opportunities for women. Women make up 40 percent of the wage-earning work force. Most of them work for the same reasons as men—to support their families, to fulfill their potential, to win admiration and respect, to express creativity. But prejudice against women leads to discrimination against them in employment. They frequently are denied management jobs. Often they are directed only toward "women's occupations," which are thought of as nursing, teaching, secretarial work, and social work. Ironically, women are treated like a minority even though they make up more than half the population.

Prejudice and discrimination also affect minorities such as blacks, Spanish-Americans, American Indians, Jews, Eastern Europeans, Italians, and Orientals. These people often are the victims of cruel jokes that attribute undesirable characteristics to their nationalities.

Prejudice and discrimination also affect the handicapped and the aged. In dealing with prejudice, remember that each human being is a unique individual. Try to understand what makes them act as they do, not prejudge how they will act.

A Difficult Situation. Prejudice and discrimination are hard to deal with. If you are the target of discrimination, it is difficult not to become defensive, but it is important not to overreact.

> Mona had a hard time at her job at the farm supply co-op. Mr. Solomon, her supervisor, made comments about how women couldn't handle tools. Mona's feelings were hurt, but she resisted lashing back. Mr. Solomon would not stop his comments even after Mona told him she didn't like them. She took her problem to the manager of the co-op. Mr. Carver said

he would look into it, but the remarks continued. Also, Mona did not get a raise when men who did the same kind of work got raises. The company lacked a grievance procedure, so Mona sought legal advice. Her lawyer told her to file a complaint with the U.S. Equal Employment Opportunity Commission. Then the harassment stopped, her raise came through, and she was placed under a different supervisor.

Jobs should not be stereotyped as women's or men's work only.

Mona's approach was not easy, but it was mature and responsible. She often felt that her human relations skills were being taxed to the limit. But by remaining calm and working through channels, she earned the respect of her employer and coworkers. And women who were hired later by the firm found the going easier.

Don't Bend Over Backward. Treating other people as if they were perfect can be as humiliating as outright discrimination and just as prejudiced. You don't need to make friends with every one.

Mr. Thatcher was a good horticulture teacher. When his school was integrated, he was determined to be fair to the new black students who came into his classroom. The new students were eager to earn the respect and admiration of their teacher and classmates. They never got a chance. Mr. Thatcher never gave them time to answer questions, but quickly prompted them. He offered them help, even if they didn't ask. He did not understand why all the students started to resent him. After observing the class one day, the principal told Mr. Thatcher: "I know you mean well with these new students. But you're not giving them a chance to show their stuff. Why don't you treat them exactly like the white students. That will show a true lack of prejudice."

After that, Mr. Thatcher was able to permit the new students to make mistakes, just like the others in his class. He found that some black students did well in horticulture, some were average, and some did poorly—just like all the other students he had ever taught.

Human Relations and You: A Review

Human relations are not obvious or easy. It takes a lot of practice and maturity to develop an effective method of dealing with people. You can't learn it in the same sense that you learn to swim, play a guitar, or deliver a speech. You need constant awareness of, and sensitivity to, other people that is grounded in your own awareness of, and respect for, yourself. If you lack this self-respect, it will be difficult to have positive human relations, no matter how hard you try.

Remember, other people have many of the same doubts and fears you have. Each person is a unique mixture of characteristics. Everyone has strengths and weaknesses. If you realize this, you can accept your weaknesses and work to improve them, emphasize your strengths, and meet the world with as much self-confidence as the next person.

THINKING IT THROUGH

1. What are human relations?
2. Why is it important to be able to get along with other people?
3. What are the basic goals of human relations?
4. Why is it important for you to "know yourself" when you seek to improve your human relations skills?
5. What are the basic needs of people that influence the way they react?
6. Why is prejudice harmful to others and to yourself?
7. How can minority groups or persons best be helped to realize their potential?

CHAPTER 6

GROUP DYNAMICS

People must be able to lead their own lives before they can perform effectively in groups. Effective participation requires human relations ability. This includes being sensitive to other people, knowing what makes them tick, understanding oneself, and working to eliminate prejudice. But you also need a good idea of how groups function. As you develop personal and leadership skills, it is natural to apply them in group situations. How can you contribute as a member? What role does leadership play in group functioning?

The simplest definition of a *group* is a gathering of three or more people. But not all gatherings are groups in the sense used here. All the people on a bus are not necessarily a group. But they would be a group if they were members of an FFA chapter.

Groups engage in human relations and communications. But the people on the bus may speak to each other and still not be a group. Something more is needed. A group also has a common purpose and goals, like FFA's goal of developing agricultural leadership, citizenship, and cooperation.

For a group to continue to exist, some of its purposes and goals must be met. People must get what they want from the group, such as developing leadership skills in FFA. The group must provide members with activities that satisfy them and meet their needs.

Both formal and informal groups operate on the basis of *group dynamics*. The groups must have a common interest, purpose, or activity to survive. That means face-to-face interaction among people in the group. In this sense, if you and your friends meet to talk about sports or farming regularly, you are a group. You have a common interest (sports or agriculture) and a common purpose and activity (to talk about your interest).

A group like that is informal. Social activities often occur in informal groups. But sometimes business is done in informal groups as well. Business executives may meet regularly at lunch to discuss common interests. Informal groups don't have a lot of rules or procedures. There may be an expectation that everyone attends meetings and contributes to discussions. Everyone agrees on why the group meets and how it operates, but nobody writes it down. Informal groups lack a constitution, officers, or any systematic way to take in new members.

It is easier to examine group dynamics in formal organizations. They include schools, businesses, or community organizations, which exist over time and have a systematic structure and purpose.

Groups work to meet a common goal.

Rules are spelled out, officers elected, and goals and projects decided on by a systematic process.

CHAPTER GOALS

In this chapter your goals are:

- To describe how groups function
- To participate in setting group goals
- To identify different member roles in a group
- To contribute to the effective functioning of a group

Group Development

A group of farmers formed the Rose Valley Improvement Association to work on soil conservation, beautification, and water management. The group developed a program to plant trees and clean out trash dumps along the highways. At first, members were excited, but the group eventually fell apart. The Bronsons wanted to work on flood and erosion control first. The Morenos wanted beautification to focus on secondary roads instead of major highways. Others wanted the group to cooperate in buying seeds and fertilizer to improve the soil on their farms. The meet-

Both formal and informal groups operate on the basis of group dynamics.

ings degenerated into arguments and members drifted away.

This group lacked the common purpose that is essential for effective group action. Members joined the group for personal reasons that they could not accommodate to group goals. A few people had established the specific goals, and those who joined the group later did not have much influence. Participation and lasting membership in groups is encouraged when all members help set group goals.

How Groups Stick Together

During the Revolutionary War, Benjamin Franklin told fellow patriots: "We must all hang together or we will all hang separately." This was a call for *group cohesiveness*. Groups stick together better when they provide services their members want and need in a cooperative atmosphere. Cohesiveness develops when groups deal with problems members think are important, provide members with recognition, promote free and friendly interaction among members, encourage cooperation, and offer favorable evaluation of member activities.

> Residents of Berry Mountain thought their tax assessments were unfair and that roads were bad in their area. When they formed the Community Improvement Committee, they focused on unfair taxes and poor roads. The issues were important enough to members to build group cohesiveness. Member recognition came in dealing with government agencies. Each person was selected to speak for the group at different times.
>
> Open interaction among members was encouraged by leaders who made sure everyone got a chance to speak. Social hours were held after each meeting so members could become better friends. They worked well together because their views were heard and respected and because they liked each other.

Leaders made sure that the more interesting jobs were passed around. The person who stuffed envelopes one day would speak at a press conference on another day. So when there was work to be done, everyone pitched in. Everyone recognized the need to cooperate to complete each task.

There was praise—for each task well done. Leaders thanked those who did a good piece of research. Public credit and praise was given to the person who came up with a good idea.

All of the factors contributed to the spirit of the group. But not all of the factors are needed all the time for effective group action. What is important is for members to feel that they belong, that they are making a contribution, and that their contribution is making a difference.

The Group Perspective

Groups do not start out "hanging together." Cohesiveness must be worked for. Groups are organized and develop cohesiveness around common goals. For example, FFA chapters are organized for students who are enrolled in vocational agriculture courses and who want to develop leadership abilities. Public Law 81-740 granted a national charter to the FFA, recognizing it as an organization for students in vocational agriculture programs.

As a member of the FFA, you will find many of the goals already set. However, you can help decide how existing goals are to be carried out or help set new ones. True group goals must be developed with interaction and cooperation from group members. If members don't accept the goals or work to change the goals, the group can lose its effectiveness.

The Mallory FFA needs a community project. The officers decide to set out trees for a local nursing home. If other members disagree, the project will fail. The trees will die, members and officers will be angry, and the chapter's reputation will suffer. But suppose the president tells the chapter: "A few of us discussed planting trees around Lakeland Nursing Home as a community service project. What does the chapter think?"

One member notes that some trees at the home are old and may need to be cut soon. Another knows how much trees will cost. A third suggests planting some trees that grow fast for quick shade and others that grow more slowly but live longer. The president picks a committee to research the idea. Members are chosen who expressed good ideas and who usually work hard. The committee reports back at the next meeting and the project is approved and the trees planted.

Not all projects are accepted this easily. Leaders and members should not push pet ideas that are not supported by the group. They must involve most members and be prepared to give up on projects that are not accepted by the group.

George pushed hard to get his FFA chapter to set up an environmental project to recycle waste paper, metal, and glass. The chapter already had wildlife management and soil conservation projects. But George insisted and the chapter gave in. Unfortunately, the project failed because members were not committed to it.

Individuals and Group Goals

Group goals evolve from the goals of members as they interact in the group. In groups like the FFA, the goals may be long-term, such as leadership development. Members join groups that have goals similar to their own. If you want to act, you would join a drama club. But if you want to learn about agriculture, you would become active in the FFA.

Group goals reflect the value system and beliefs of individuals. Students in vocational agriculture also become members of FFA because their goals are to learn about agricultural careers. You bring your own values to a group and perhaps fit them to group goals. If you think working with people is important, you might take part in an FFA community service project such as voter registration in an election year.

Establishing Goals. Each organization should have a procedure for setting its goals. In the FFA, each chapter develops a program of activities each year. It includes a listing of activities, goals, and ways and means to meet the goals. Effective planning must be a group process to represent the goals of the members. Here are the steps:

In planning for this year's activities, the chapter members would first describe every activity in writing. Community service would include providing services for individuals and leaving something lasting in the community.

The second step would be to list the goals for the activity. One goal of the community service program would be to develop and maintain a community park.

Third, the members would describe the ways and means to reach the goal. These could include securing land for the park, planning its layout, and developing the park facilities. It also would include the things that needed to be done to maintain the park.

Group goals can evolve from the ideas and interests of one member.

Roles of Group Members

Members must fill many different roles if a group is to reach its goals. The interaction of people in different roles ensures the success of a group. Each role carries its share of leadership potential and responsibility. Here are different roles that are open to you in group situations:

The FFA chapter has met. The *presiding officer* fills an important role. This almost always is an elected, official leader, in this case, the president of the chapter. The presiding officer runs the meeting in an orderly fashion and makes sure everyone gets a chance to speak. The president says: "The next item on the agenda is buying a tractor to help us till the school's land laboratory."

The *initiator* takes the lead and provides ideas for the group to act on: "I move that we buy the tractor."

The *recorder*, usually the secretary, keeps the formal record of a group. Following a second, the recorder writes down the motion.

The *opinion seeker* asks questions so others will express themselves: "How do other members feel about spending this much money?"

Opinion givers make and express judgments. "I think it's important to buy a tractor," says one. Another says, "I disagree, a tractor is too expensive."

The *information seeker* brings out the facts: "First, we need to know how much it will cost. Also, what other equipment will be needed and how much that will cost."

The *information giver* provides facts: "It depends on the tractor. I saw a used one advertised for $800. It was old, but was supposed to be in good condition. A new tractor would cost several thousand dollars."

The *elaborater* clarifies issues and adds details: "We probably would need a plow and harrow. We could borrow equipment from farmers when they weren't using it."

The *critic* questions ideas and points out problems: "Do we need a tractor? It will be more expensive than we think. We'll need insurance to cover members who operate the tractor and gas and oil. There also will be repair costs."

The *summarizer* clarifies the group's position and states the pros and cons: "This is not a new idea. It was discussed last year. The chapter rented land 2 years ago to give us practical experience with crops and soils. Farmers have loaned us equipment, but we have a hard time getting tractors. Farmers use their tractors almost every day, but not the other equipment. So we may not need to buy extra equipment."

The *integrator* coordinates and relates ideas and activities: "This tractor wouldn't be used only on our field. People in town have a hard time getting their gardens tilled and snow removed. The tractor could make some money and provide a community service."

The *expediter* tries to get the group to act by preparing, providing, and passing out materials: "Here is an analysis of the costs of various ages, sizes, and models of tractors; the cost of equipment; and the cost of operations and maintenance. Take a copy and pass the papers around."

The group's ideas are praised and stimulated by the *encourager*: "This has been a good discussion. Members have done a lot of homework on this issue. But we need to know more. Let's have more discussion of costs in relation to the chapter's other projects. The critic made some good points. What will repairs cost?"

The *harmonizer* tries to reconcile differences: "This question has been debated in either-or terms. Maybe we can buy a tractor and still have money for other projects. Other chapters have tractors and other projects. They hold down repair costs by careful operation of the tractors."

The *standard setter* expresses standards for the group: "Let's remember that we're here to learn to be effective workers in agriculture. That experimental field has been important to us and the school. Vocational agriculture students say work in the field has improved their classwork. A tractor might let us do more projects instead of less."

It's Up to You. You will likely fill many of these roles as you work with groups. They all are important in places where you may work, such as school organizations and businesses. If you can assume some of the roles, you will make important contributions to any group. You may fill one role or more for the chapter meeting and others for a committee meeting. One time, you will want to draw out opinions of group members. Other times, you will want to express your own opinions or share information.

On the job, you will act as an opinion and information seeker many times. As you gain experience, you can become an initiator, elaborater, harmonizer, or standard setter. Each role is an important aspect of leadership. Only the presiding officer and recorder are formal leaders. But the other roles provide opportunities for, and show the importance of, informal leadership. Only your own knowledge, imagination, energy, and initiative will limit the number of different roles you can fill. Each role makes a different kind of contribution, but they go together to make up a harmonious whole. Both the critic and harmonizer are needed.

Group Leadership

A good atmosphere is essential for successful operation of a group. Effective leadership is needed to create a positive atmosphere. If an FFA president runs meetings efficiently and follows proper parliamentary procedure, the chapter will

The treasurer of an FFA chapter may fill many roles in the group.

be, and will seem, more effective. Decisions will be made without bickering or divisive arguments. This means a pleasant atmosphere in which members are eager to offer their opinions. Poorly run meetings make it hard to gain recognition and to be heard. An unpleasant atmosphere discourages members from attending meetings.

Both leaders and members are needed. Dynamic, imaginative leadership can do little if the group is apathetic. But ineffective leadership can dampen enthusiasm. Leadership should be the spark that ignites a group to work to achieve its goals.

The Best Kind of Leadership

A democratic atmosphere and approach promotes dynamic interaction between leaders and members. Members decide the group's goals and programs. Leaders and members are committed to seeing that group decisions are carried out.

Democratic leadership is most effective at arriving at better decisions in groups in which membership is voluntary. It also is superior in building group cohesiveness and promoting cooperation. *Autocratic* (leader dominated) and *laissez faire* (little leadership) groups do not cooperate and interact with a purpose. Under autocratic leadership, members do not influence the group's actions and care little about seeing the actions succeed. Laissez faire leadership gets in the way of effective group interaction because "nobody's minding the store." Without effective leadership, groups can't focus on goals or follow through.

> The Western Grebe Society was a laissez faire group dedicated to wildlife protection. Members believed that strong leadership stifled democracy. Everybody got a chance to speak, but that was all. No one wanted or knew how to enforce parliamentary procedure. There was no way to end debate or to bring questions to a vote. The presiding officer did not require members to stick to the subject. The society did little to protect wildlife. While members of the original group debated, a group broke off, formed a new group with rules and procedures, elected strong but democratic leaders, and persuaded the county government to establish a bird sanctuary in the park.

Democratic leaders can influence who works together. This can be done informally: "George, will you and Marie get together and write the news release by tomorrow?" Or the leader can be more formal and appoint a committee: "Tony, Donna, Martin, and Lucy, will you serve on the committee and report back to us in two weeks?"

Promoting Group Functioning

A skillful leader develops many ways to get groups to interact. The leader needs to plan to make sure that members participate productively. For a meeting, the leader should have the objectives firmly in mind and be able to state them clearly: "Today, we elect officers for next year." The leader schedules time for discussion of each objective, issue, or motion that comes up at the meeting. Time is important. Start meetings on time and establish a closing time. Limiting time encourages members to stick to business. Definite times also permit members to plan their schedules and stick to them.

Inform members of the agenda before meetings, so they can think ahead about how they want to participate. The room in which the meeting is held should be comfortable, well ventilated, and well lighted to create a pleasant atmosphere. And don't hold a small meeting in an auditorium or a large meeting in an office. Make sure that the space fits the group.

Group decisions should be followed by committee appointments or activities. If your FFA chapter votes to take a field trip, appoint the committee to plan the trip as soon after the decision as possible. This shows members their opinions and decisions are respected. Ask the committee to report back at the next meeting, even if the trip is several months off. This helps move ahead on the project and gives the truthful appearance of effective action.

Encouraging Group Interaction

Effective leaders plan the kinds of interaction they want for the situation or the goal they want to accomplish. Here are some techniques for promoting interaction:

Buzz Groups. Everyone wants to speak at once at some meetings. Break the larger group into groups of six or eight people. Name a leader and recorder for each group. This technique has three advantages. First, the charge is to examine as many ideas as possible and report back. Second, all members will get a chance to speak in the small groups. Third, more members can practice their leadership skills. The group leader and recorder are formal leaders. And each member can try out some roles such as initiator, information seeker, or critic.

Panel Sessions. This is a good way to present expert information. Panel sessions can take several directions:

A *question-and-answer session* gives members a chance to ask panel members specific questions. A panel might include a banker, an insurance agent, and an accountant. Members could ask the banker about different types of savings accounts in which to deposit chapter income.

In a *symposium*, each panel member speaks in an area of expertise. The accountant discusses setting up a bookkeeping system. Then members can ask questions.

In a *problem-solving session*, panel members deal with specific problems. The insurance agent could discuss how much and what kind of insurance is needed for a tractor.

Brainstorming. How can your FFA chapter raise enough money to pay for all of its projects? You ask the group to brainstorm. Members state every idea that comes into their heads without criticism. After every idea has been stated, then they all can be evaluated.

Questioning. Questions stimulate group interaction. They help people generate ideas and promote thinking: "Do we really need a tractor? Is there a better way to do it?" Questions may be used to involve quiet members: "Mary, you know a lot about gardening. How many tomato plants do we need to have enough tomatoes for our canning project?" Questions can summarize a discussion or move a group to action: "Have we tied up all the details for the farm fair?" "Now, who will pitch in and clean up after the fair?" And questions may be used to check an agreement: "Do we agree to finish the project before the new year?"

Leaders must encourage group function and interaction.

Getting the Job Done

One leadership role is to encourage positive performance by members. *Performance* is positive action that affects the group or that makes a member relate to the group. The performance of individual members determines the effectiveness of groups. When members interact, help set goals, or take part in activities, they are engaged in positive group performance.

> Nancy rarely spoke out at meetings and people did not think of her as a performer. But her performance was important to the success of the FFA chapter. She took responsibility for all paper work. Nancy always volunteered to write letters, type reports, or do mimeographing. She knew the importance of those jobs and enjoyed them. Her performance was positive.

When seeking to involve members, leaders should use the interests and abilities of each member.

Ray was quiet at FFA meetings. He rarely performed. The president knew Ray was a whiz at math and liked it. He asked Ray to serve on the finance committee. Ray's interest in all chapter activities picked up. He helped the treasurer set up a better bookkeeping system and spoke with authority on financial matters.

Sometimes, getting the job done is a slow process. Leaders can become frustrated and discouraged if they don't allow for the time it takes to make the democratic process work. But the sounder decisions and stronger groups that result from operating democratically make the effort worthwhile. The process is often as important as the decision when you are trying to build a strong, active membership.

Nobody can force members to become active, but leaders can encourage participation. An apathetic member, if encouraged, can become an important source of energy in a group. This new energy makes the group more active and vital. One excited member can excite others.

How Groups Work: A Review

Every member doesn't have to be excited about every project. Some will focus on certain goals and go along with others. One FFA member will be interested in community service, but not as interested in leadership development. Another may take an active part in leadership activities, but pay less attention to agricultural projects. But members who are excited about one part of a group's work will be willing to help on other parts. The secret of a vital and successful group is to encourage all members to make the group theirs. If they share participation and pride in the group, they will want to work hard. And they will have fun doing so.

THINKING IT THROUGH

1. What is a group? Why do groups exist?
2. What is group cohesiveness? How can group cohesiveness be developed?
3. Why do individuals join groups? List some groups to which you belong.
4. How can members help identify goals for the group?
5. What roles do group members fill?
6. What can a leader do to promote effective group functioning?
7. How can a leader promote group interaction?

CHAPTER 7

PERSONAL COMMUNICATIONS

"Nancy, would you please get the window display ready for the spring season?" asked Mr. Gregg, the Garden Center manager. "Sure," Nancy said. But she wanted to finish stocking shelves and put off the new task. Next morning, Mr. Gregg was angry. He had wanted the display prepared right away. Nancy had not understood his message in the way that he intended.

Neither of them had learned effective communication skills. Communication is a key skill for leading your own life or leading others. Everyone needs to learn to *communicate* effectively. This means sending *and* receiving messages. Communication is the most important aspect of human relations. When you communicate, you have human relations. Mr. Gregg did not communicate his message to Nancy clearly and had poor human relations.

However, communication and human relations are not identical. You can have human relations and not communicate. If you work on an assembly line that makes tractors, there may be so much noise that communication is impossible. But you are cooperating with others as you work, so you are having human relations.

For effective communication, the message must be sent and received. A survivor in a lifeboat who sends an SOS message on a radio frequency to which nobody is tuned is not communicating. The message is not being received. But if the survivor waves a torn shirt and a ship sees the signal, very effective communication occurs.

Messages must be conveyed accurately. If the receiver understands a message differently from the sender, this may be worse than no communication at all, as shown in the example of Nancy and Mr. Gregg. Effective communication means that the listener understands what you intend to say. It also means that you say clearly what you intend. The ability to speak and write clearly and to listen and read carefully are essential for effective communication.

To have good human relations, you must have the ability to communicate clearly. You send and receive messages all the time. Teachers communicate lessons, instructions, information, and test questions. If you receive these communications inaccurately, you will have trouble in school. Supervisors at work send and receive messages. If you don't understand a message, it is up to you to clarify it. Ask for an explanation, or repeat the message back to make sure you understand it.

CHAPTER GOALS

In this chapter your goals are:

- To initiate a conversation with another person
- To analyze key clues to nonverbal communication
- To follow the correct procedure when using the telephone
- To make introductions in different situations
- To write business letters
- To evaluate the importance of effective communication in developing leadership skills

Oral communication is an important aspect of human relations.

Oral Communication

Conversation is the most common form of oral communication. You have conversations when you meet a friend for a snack, when you chat between classes, or when you talk to someone on a bus or airplane. *Conversation* usually is defined as an oral, usually informal, and pleasant exchange of ideas.

How often have you sat next to someone on a bus or at a meeting and wondered who the person was? How would the person react if you spoke? What would you say? The other person seemed pleasant, but nothing happened. How do you open a conversation? Somebody has to begin.

Striking Up a Conversation

There are several factors to consider before striking up a conversation with a stranger. They determine the kind of opening remarks you make, how long you talk, or even whether you talk at all. The factors include where you are, who the other person is, when you should speak, and what to talk about.

Where Are You? The location and situation are important. Obviously, you must be able to hear the other person. Conversation would be difficult near a noisy jet runway. You also should make sure that you don't inconvenience others. The place should be appropriate for talking. Just after an FFA meeting starts is not the right time to start a conversation. However, a quick, whispered introduction might be in order.

Also, are you on home ground or in strange territory? If you are in a familiar setting and the other person is not, it is appropriate and courteous to introduce yourself and start a conversation. The person may feel ill at ease and want to meet someone. On a new job, you might wait a little to give other employees a chance to take the initiative.

Whom Are You Meeting? Consider the other person. Do you seem to have common interests? If so, opening a conversation is worth a try. Or both of you might just want to pass the time pleasantly in the company of another person. This can occur at work during breaks. You can make friends, have a more pleasant lunch, and learn something about the business in conversations with coworkers. If you work with someone, there is a chance you have common interests and may develop a friendship.

In sizing up other people, be alert for clues that can provide you an opening. Notice the books, newspapers, or magazines they are carrying: "I see you're reading the *National Livestock Producer*. I have a small herd of Angus." They may be carrying tennis, skiing, or other sporting equipment. Look for rings, pins, or emblems that give a clue to their interests or professions. An FFA pin is enough to provide an opening: "I see you're an FFA member. So am I. My name is Mary Knowlton."

If you have been introduced, it is always appropriate to talk, at least briefly. The introduction may include information that can lead to further conversation: "This is Harold Powers. He was regional star farmer last year."

When Do You Talk? Does the other person seem receptive? Is the person idle or busy? Friendly or forbidding? Don't interrupt a person who is reading intently or speaking with another. People often work when they are traveling. They read, make notes, or plan for the meetings they are attending. They will not appreciate an interruption.

> To Manuel, the situation seemed opportune. His seatmate was not busy and seemed friendly. It was worth a try. "Are you going to Chicago too?" he asked. "Yes," smiled his companion. "I have business there. Why are you going?"

What Do You Say? This also depends on the situation, but a number of topics can open a conversation. Several clues are given under "Whom Are You Meeting?" If you don't pick up an idea from the person, weather can be used as an "icebreaker." It is familiar to everyone and not intimidating. While chatting about how hot it has been, you can size up the other person's desire to continue the conversation. Is the person interested or merely polite? What would be your reaction to the following situation?

> You say: "It sure has been hot lately."
> Your companion says: "Yes, crops will be hurt if we don't get some rain. My corn needs rain badly."

Your companion is probably ready for more talk. If the response is a nod or a "Yep," you know the person probably prefers silence. Topics that are familiar, that do not threaten, and that are of common interest provide conversation starters. The weather is particularly appropriate for people who you know are interested in agriculture because it is important as well as familiar. Crops, livestock, or markets also are good topics for such people. Other conversation openers can be sports, school events, travel destinations, business concerns, music, or anything of general interest.

Until you know another person, avoid

Before starting up a conversation, make sure that your message can be clearly understood.

topics that are controversial or that might be offensive. They can end conversations quickly. The person you want to meet is reading a newspaper. When the newspaper is put aside, you say: "Candidate Y is a real dummy, don't you think?" Your companion strongly supports the candidate and coldly informs you of that fact. End of conversation and chance for friendship.

Don't push a topic that only interests you. Be alert to signals from the other person. Some people may politely let you ramble on about your job. If you are doing most of the talking, the other person may be merely polite—and bored.

Plan some topics you can use in a variety of situations. For example, a new student comes to your school. You are sitting next to the featured speaker at the FFA banquet. You are on a date with someone for the first time. You run into your boss in line at the bank.

Effective Listening

Talking is only half of oral communication. The ability to listen carefully is another key to effective communication. Some people spend the time the other person is talking in thinking about what to say next. They don't pay attention to what is being said. The effective communicator listens carefully. If you do so, your response will be on target and will make the other person want to continue to talk to you.

Effective listening is a signal that you want to hear what the other person is saying. It shows courtesy and respect for the person speaking. A friend is talking to you. You're not paying attention, but want to seem to be in the conversation. The friend says, "I did my best, but I got to the meeting late." You tune in on the last word, but don't get it quite right: "Which lake was the meeting held at?" you ask.

Developing Listening Skills. You can develop your listening skills. At first, it requires an effort to keep your attention focused. But with practice, good listening becomes automatic. Keep eye contact with the other person. The animation of the eyes and face will help you stay interested. It also shows that you are paying attention.

Respond to the other person's comments with an occasional word. But be careful with nervous conversational habits. Words such as *yeah* or *really* can be annoying if they are repeated constantly. You can break these habits if you become aware of them.

Questions are another way to show your interest. There may be something you don't understand or more information you would like to know. You show your interest by asking questions that focus on the topic of conversation. If you are paying attention, your questions will be appropriate, not ones that are likely to be answered in the next sentence or two.

> Ralph: "Plowing my bottom land was really tough."
> Dick: "Was it too wet?"

Dick was trying to indicate his interest with a question. But Ralph probably would have given the information in the next sentence. The next example shows a more appropriate question:

> Harry: "Mr. Kensington has a really fine dairy herd. Its milk production is tops in the country."
> Luann: "Does that include butterfat production?"

Luann's question indicates interest and seeks information that Harry might not have given in the course of the conversation.

A simple question may begin or continue a conversation: "Hi, I'm Mark Strand. I see you're wearing a Mallory High School ring. When did you go there?" Questions indicate an interest in other people and their opinions. But avoid asking for personal data: "How old are you?" "How much do you make?" Such questions are best left out of casual conversations. But positive responses may be given to questions like these: "Where are you going?" "Do you like bluegrass music?"

Questions also can be used to clarify remarks: "How long did it take you to complete the experiment?" Or they can ensure that you understand what the speaker wants: "You mean I should lubricate both tractors?"

Body Language

You communicate constantly whether you are speaking or not. The term *body language* means that you send messages with your whole body, not just your mouth. This is *nonverbal communication*, the sending of messages that are not spoken or written. Often, the only clue to what a person is thinking or feeling is a gesture of the hand, a facial expression, or the position of the body. All of these things communicate thoughts or attitudes.

You can learn to read the nonverbal messages others send you and become aware of your own. Sometimes, people don't know they are sending nonverbal messages. At other times, these messages can be confusing.

> Allen invited Cindy for a talk. But he kept looking out the window. Cindy thought Allen was bored and prepared to excuse herself and leave. Then Allen said: "My brother asked me to watch for the mail. He's expecting a package and doesn't want it left outside." Cindy had misread Allen's nonverbal message.He could have avoided causing her discomfort by telling her about the situation when she arrived.

Many common expressions are clues to nonverbal messages. "Her face is like an open book" means she communicates her thoughts and feelings through her facial expressions. Facial expressions communicate a lot. If you can't read someone's face, the message may be: "I guard my thoughts and feelings." A smile may mean acceptance, friendliness, agreement, or pleasure. But some people smile out of habit, and their smiles don't mean much.

Reading Eyebrows. A frown may indicate thought, disagreement, concentra-

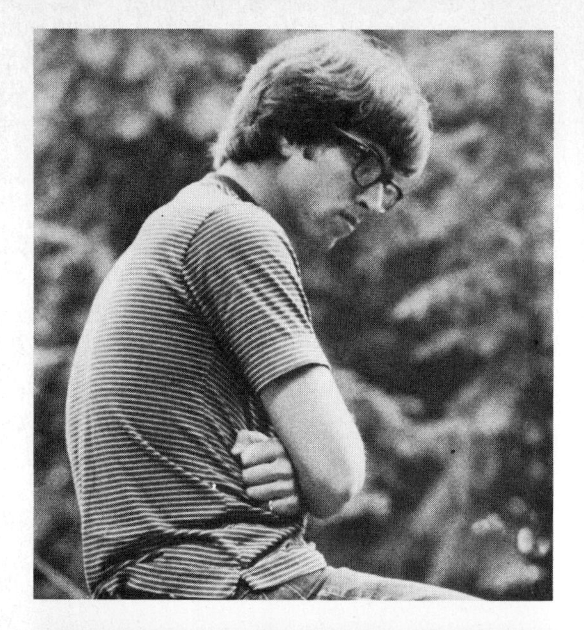

Body language is a form of nonverbal communication.

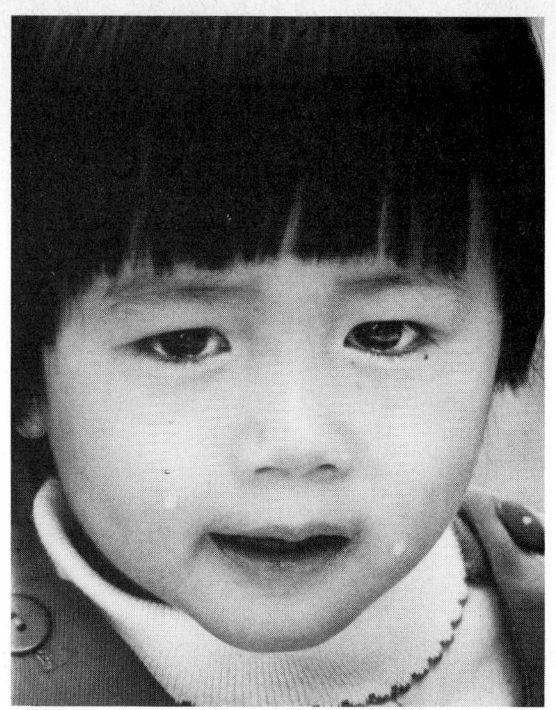

tion, or puzzlement. It can also indicate discomfort or nervousness. People's eyebrows tell a lot. When they are pulled together, the person may be puzzled. When both are raised, the person is probably surprised or frightened. One raised brow may indicate disbelief. The jaw also sends messages. If it is firmly set, it may signal controlled anger, pain, or sadness. Anger is indicated in the expression "He spoke through clenched teeth."

"They greeted me with open arms" indicates welcome. The body and hands also express thoughts nonverbally. The clenched fist can show anger or determination. Tightly folded arms may show determination or stubbornness. To slouch, yawn, or rest your head in your hands may send a message that you are bored or discouraged.

Using the Telephone

Telephone communications are becoming increasingly important. The telephone is used extensively for business and social occasions. The business executive may use the phone to make an appointment, set up a meeting, or work on a deal. The phone may be the only contact a business has with much of its public. Socially, the phone can be used to issue an invitation, arrange a date, or express thanks. But the telephone limits communication to the spoken word. Nonverbal clues cannot be seen.

Your voice carries the message. Messages have both a verbal and emotional content. Nonverbal clues help establish the emotional content. But the voice also expresses emotion. In talking on the phone, try to keep your voice cheerful and pleasant. A gruff voice can turn the caller off to you or the group you represent. Speak distinctly and clearly. The other person can't see your lips and expression, which aid in understanding direct oral communication. Make sure you use a lively voice that expresses alertness and interest. Effective telephone communication is similar to the face-to-face kind. Keep your attention on the message and create a feeling of confidence.

> Norman's office phone rang. He picked it up and said: "Yeah.... I dunno.... Naw, he's out.... Good-bye." His telephone style needs much improvement. The caller had no idea who answered the phone and probably took his business elsewhere. Norman made no effort to be helpful or to take a message. He needs an immediate lesson in effective telephone manners and techniques.
>
> Just think how much better the result would have been if Norman had answered the phone and said: "Boone Feed Supply, Norman speaking.... I'm sorry, Mr. James is at the warehouse. Can I take a message?... Thank you. Good-bye."

Speaking Techniques. The other party hears best when lips are held about ¾ inch from the phone. Pronounce names carefully and spell them if necessary. Even the so-called common names contain some surprises: *Smith* may be spelled *Smyth*. The name may be *Johnson* or *Johnston*. *Brown* and *Green* may be spelled with an *e* on the end.

Complete sentences help the other person get an accurate sense of what you are saying: "Ms. Jones is out to lunch." "No, we don't stock that brand anymore." In addition, organize your thoughts before speaking, and don't interrupt the other person.

Making Introductions

An insurance agent comes to an FFA meeting to talk about the different kinds of insurance needed in agriculture. There are a few minutes before the meeting to introduce the guest to the FFA officers. The use of standard introductions creates a comfortable situation for all concerned.

You make introductions to provide names and brief information about persons who have not met. "Mr. Rivera, this is Joe Padgett, our secretary. He's interested in dairying. Joe, this is Mr. Rivera, our guest speaker tonight. He's president of Rivera and Gomez Insurance Company."

Situations vary. You might use a slightly different approach at a meeting, in a business situation, or at a social occasion. Different numbers of people also may be involved. You are introducing yourself to another person or to a group: "Hi, I'm Thomas Pulaski. I'm the representative from the Columbia High FFA." Or you may be introducing a person to another person or to a small group.

Introducing Yourself. It is usually customary to offer your right hand and identify yourself: "I'm Peter Greenberg." Either you or the other person can initiate an introduction. Most people are complimented when another person wants to know them. Your introduction of yourself can be accompanied by a brief explanation of who you are: "I work for Golden Leghorn Farm."

Introducing Others. There are accepted ways to introduce people to each other. Younger people are presented to older persons: "Mr. Agostino, please meet Alden Pruett." Etiquette calls for presenting the man to the woman, even if she is younger: "Ms. Willoughby, please meet Hank Strouther, my uncle." Usually, the name of the woman, the older person, the stranger, or the notable person is spoken

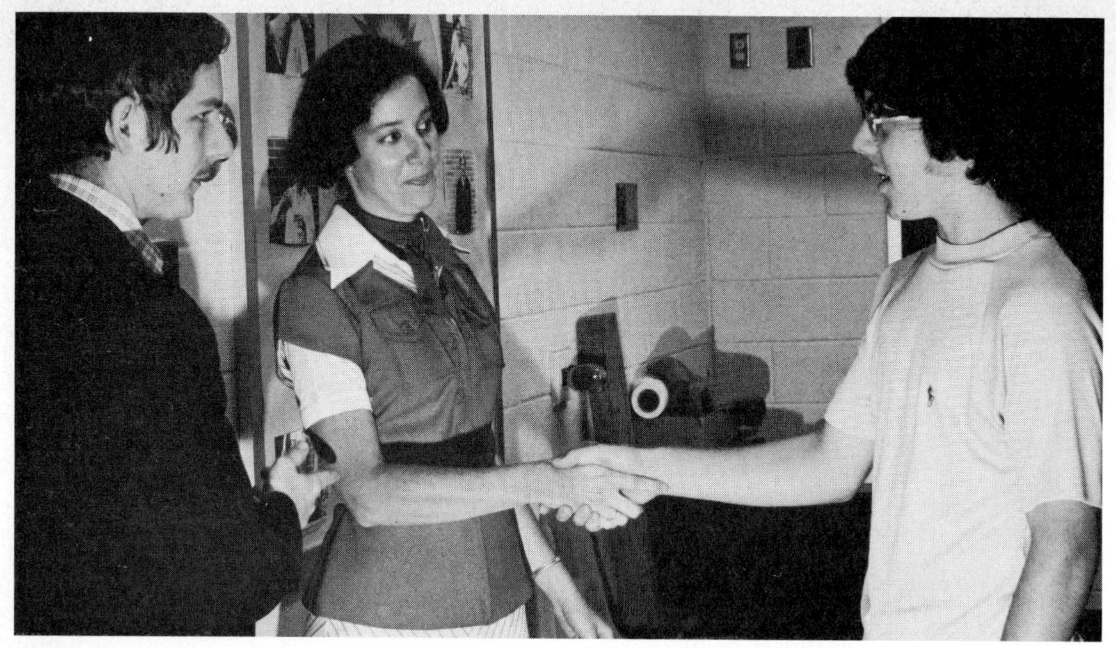

There are certain customary ways of introducing people.

first: "Dr. Banks, please meet Lewis Striker." Then add "Lewis, Dr. Banks," to complete the introduction.

You also will introduce people to groups. You must decide if the person should meet everyone in the group. If everyone stops to meet a newcomer, then provide introductions. If there are three or four people in the group, introduce everyone. But for larger groups, introduce the few people who are seated or standing near the visitor. Always introduce visitors to officers of your organization. Use a simple form of introduction: "Hello everyone. This is Betty Lennon." Then, indicating with gestures: "Betty, Sue Jones, Tim Garver, Max Trueblood."

Writing Letters

There are many forms of written communication and many different occasions for writing. It pays to develop good letter-writing skills. A letter may be your first, and maybe only, introduction to another person. It can make a strong impression. If you are sending a resume to a possible employer, its appearance will make as much of an impression as its contents.

In business, letters may be the only chance to make a sale. Messages should be written clearly with correct grammar, spelling, and punctuation. Improper grammar quickly loses the attention of readers. The words and the condition of the page carry the whole message. There are no nonverbal clues or tones of voice. You need to state your feeling more explicitly: "It was good to hear from you." "Thank you again for the interview."

Reasons for Writing Letters

Letters are used for various reasons. They permit communication with people who live too far away to call inexpensively. Letters often have a specific business or social purpose. They may provide specific directions or details. If you are inviting a speaker to address your FFA chapter, you need to mention the date, place, time, name of your group and school, and directions on how to get to the meeting. You may want to suggest a topic for the speech. A farm manager might be asked to speak on crop production. Be specific and include all information.

Letters also provide a written record of agreements, arrangements, or specific information: "How much does the harvester cost?" "What are the rules for the livestock-judging contest?"

Importance of Good Letters. The appearance and content of your letter determine the impression you will make on the reader. Neatness, correctness, and good spelling and grammar are essential. The reader makes a judgment about you based on your letter: "Say, this person is neat and courteous. I could use someone like this as a sales representative." If the letter is sloppy or contains a lot of mistakes, it conveys another impression: "This is not the sort of person I want to meet my new customers."

A very important aspect of letter writing is the organization of what you want to say. The topic of the letter may be introduced briefly: "I want to apply for the job of forester that was advertised in the April *Forestry Magazine*." Then discuss as thoroughly as necessary what else needs to be said: "I have studied forestry in high school. I worked for Harding Lumber Company for three summers, both in the mill and cruising in the forest." If necessary, close the letter with a brief summary that pulls together the thoughts you have expressed. "I believe I am qualified for the job and I certainly am interested in it."

Writing a Good Letter. Every letter has certain basic parts. Business letters are more formal and have more parts than personal notes. But even personal letters have a basic form that depends on who will receive them. A thank-you note to a relative will be a little more formal than a chatty letter to a friend. The illustration above indicates where to place the following basic parts of a letter:

May 23, 1976

Mr. Glen Erickson
Manager
Williamson Farm Cooperative
P. O. Box 445
Williamson, Wis.

Dear Mr. Erickson:

I want to apply for the opening on the sales staff that was advertised in the May 2 issue of the Williamson Chronicle. Enclosed is my personal data sheet. You will notice that some of my experience is related to your needs.

I would appreciate an interview at your convenience.

Sincerely,
Drew Hammond
Drew Hammond
Box 129
Williamson, Wis.
941-3265

A sample job application letter.

- Address of writer.
- Date letter is written.
- Name and complete address of the person to whom the letter is written.
- Salutation: "Dear Mr. Walker" or "Dear Claire." If you know the person well, use the first name; otherwise be more formal.
- Body of the letter, organized as follows:
  Purpose of letter: Applying for a job.
  Facts or requests for information organized into paragraphs: Your qualifications. A request for more details about the job in question.
  A concluding statement: Restate your interest in the job and your confidence that you can handle it.

- Closing: Phrases such as "Sincerely yours," "Sincerely," or "Cordially" are used most often, especially in business correspondence.
- Your name: A signature. If the letter is typed, your name should be typed below the signature, which is in ink.
- Your title or office, if you are writing a business letter.

Type business letters on plain white paper or business letterhead. Use even margins on both sides and at the top and bottom. Regular spacing of words and lines improves appearance. Never use notebook or lined paper for letters. Handwritten letters are acceptable for personal communication and generally are expected for letters of invitation, acceptance, regret, congratulations, appreciation, or condolence.

Handwritten letters should be neat and readable. Use a pen with blue or black ink. Form letters carefully. Use the same format as in typed letters. Make sure that all words are spelled correctly. When in doubt, use the dictionary.

Fold the letter so it opens ready to read. Make as few folds as possible to fit the envelope. Place a return address on the envelope on the upper left corner. To address the envelope, use the complete address, including the title of the person receiving the letter. In all written communication, carefully read what you have written and correct all errors. Read business letters more than once for errors.

Communicating Effectively: A Review

Effective living requires the ability to communicate effectively. Everyone needs communications skills for business, school, and social activities. Everyone has looked at the personal aspect of communication, which everyone uses daily with families, with friends, in school, and at work. Like other aspects of leadership and personal development, communication skills can be learned. You can study effective communication in books or by watching those who already have acquired the skills. Most of your life is spent communicating. Only *you* can ensure that you do it well.

THINKING IT THROUGH

1. What is communication?
2. How might a person begin a conversation?
3. Why is listening an important part of communicating with others?
4. What are some nonverbal clues to communication? What can these clues tell a person?
5. Why and how can telephones be used for communication?
6. What simple rules should be followed in introducing persons to each other?
7. Why are good letters important?
8. What are the parts of a letter?

CHAPTER 8

GROUP COMMUNICATIONS

Arthur was popular at school. He excelled in athletics and had many friends. His conversations were always interesting. Arthur seemed the obvious choice for class president and was elected easily. His friends were surprised at the first meeting. He was fidgety and stumbled a lot when he spoke.

Many people are like Arthur. They communicate effectively with individuals, but freeze up in groups. This severely limits their ability as leaders. It also handicaps them on the job, in community work, or in social situations. The ability to speak easily to groups is needed constantly at school, at work, or at home. You need group communication skills to make a class report on soil types or to discuss why your FFA chapter should take part in a tree-planting project.

The ability to communicate in groups can be learned. It must be learned by people who want interesting and rewarding careers or who want to take part in community affairs. Positions of responsibility, pay increases, and promotions go to those who can speak effectively in groups. These skills are essential for advancement in business, politics, or community service.

Effective group communication is similar to effective communication with an individual. There is a speaker with a message and several people to hear and understand the message. The understanding is vital. To ensure that groups of people understand you, you need to develop your ability to communicate. Different situations require different kinds of communication. A discussion requires different preparation than a formal speech. The purpose and type of communication will determine your approach. Do you want to inform, to challenge, or to entertain?

If you provide your FFA chapter with information about a planned field trip to an agricultural museum, you use a different approach than if you are trying to persuade the chapter to take the trip. If the members of the group already plan to make the trip, they want to know about time, place, and mode of transportation. If you are persuading, you will explain why the museum is interesting. What can be seen and learned there?

CHAPTER GOALS

In this chapter your goals are:

- To take part in group discussions
- To lead a group discussion for 15 minutes
- To speak extemporaneously to a group
- To answer questions before a group
- To introduce a speaker
- To organize and deliver a 6-minute prepared talk

Group Discussions

Discussions are the way groups reach decisions. People talk over various choices, problems, and benefits of alternative courses of action. Most groups do their business through discussions. Discussions are *organized* oral communications that are designed to answer questions, solve common problems, reach decisions, clear up difficulties, or arrive at the truth. They can be a debate with two opposing sides. Or discussions can be friendly and cooperative, with everyone working toward a common end.

If you and a friend are trying to decide what to do for a weekend, you can discuss it without debating it. You might agree to backpack in the mountains. You would discuss the possible route, what to take along, where to camp, whom to inform of your plans. You would be working out the details of the hike.

Discussions have a different purpose from conversations. Discussions have a purpose other than pleasure. The topic could be the same in both cases. Your FFA chapter visits a dairy farm. You would talk about what you enjoyed. If you discussed the trip in class, you would be trying to increase your understanding of dairying.

Discussions often occur in groups. For example, your FFA committee is meeting to discuss what projects the chapter should have in the coming year. Or a farm family discusses whether to plant corn or soybeans this year.

Discussions may be formal or informal. Informal discussions involve two or more people who are trying to achieve something, but without much structure or rules to help them. Formal discussions usually have a leader, take place in a larger group, and are guided by rules and procedures. A meeting of an animal science class or an FFA chapter would involve a group discussion.

The topic of a group discussion has meaning for all who are taking part. They work together to arrive at a decision or solution. What is the best way to control corn blight? Does our firm need new suppliers of fruits and vegetables?

Effective Discussions

Discussions can get the job done or wander for hours with no solution. Like other tasks, discussions are most productive if there has been advance planning and a clearly defined process for carrying them on.

First, clearly understand the topic and keep your remarks related to it. Some people want to speak whether they understand the topic or not. This confuses the issue and takes unnecessary time. For example, fund raising for a specific event is different from fund raising in general. And to talk about people who are late with their dues would be out of order.

By concentrating on presenting information, you help focus on the topic. To talk about raffles when you are holding a fruit sale is not appropriate. In this case, information could include how many people bought fruit last year, what items sold best, and how many volunteers were needed.

Personal encounters or remarks about personalities quickly lead to deterioration of the discussion. This is called *ad hominem argument*, which means attacking the person instead of the idea. "Watch out for John's ideas. The sale was a flop last year when he was in charge."

Think ahead and plan your remarks before you speak. And remember that other speakers have information that is worth hearing. Keep your mind open to new ideas. Try to keep the discussion on track. Your point about the weather at the time of year the event is planned may have to wait until the group settles the question

In a noneffective discussion, a leader has no control. In an effective discussion, a leader recognizes speakers and has an organized agenda.

of the admission fee. End the discussion with a concise statement of agreement: "We have decided to charge $2.50 for admission" is an example of a decision to act. Or if agreement is not reached: "We still seem far apart. We agree to continue the discussion at the next meeting."

Leading a Discussion

Most discussions have leaders. Elected officers help lead formal discussions of organizations. The president usually is in charge of the meeting. And informal leaders help clarify issues, present information, stimulate discussion, and seek a solution. Even informal discussions often have leaders who ask questions, challenge, and make suggestions.

The leader is responsible for keeping the discussion on track. Is Mary presenting information or beginning to wander off the subject? Is it time to try to wrap up the discussion and move toward a vote? Can anyone suggest a compromise solution? Leaders often start the discussion: "The chapter's funds are quite low; who has some fund-raising ideas?" The leader outlines the purpose of the meeting or sets a task or goal for the group: "We decided last meeting that most of today will be spent planning our display for the state convention."

Leaders start or keep discussions going by asking questions or stating a topic for consideration: "We could plant pine trees on that hillside." "Are there any other ways to control erosion?"

Statements help focus on, or redirect group comments to, the topic: "We are considering the best route from Medway to Calverton." "We need to understand all the possible consequences of buying a tractor.'' Or the direct approach may work best: ''Please stick to the subject.''

The leader is responsible for squelching ridicule or snickers at people who are shy or just beginning to learn to speak in public. Calling on members encourages participation. Shyness before groups can be overcome. A leader can help with a nod or expression of encouragement: ''That's a good point, Sarah.'' ''Gene, you really did your homework.''

Public Speaking

There are many group situations in which one person speaks to the group. Effective public speaking is an important skill. At some time, most people have to stand before a group and share information. If you have an idea for a new sales campaign, you will want to share it with your employer and coworkers. Some situations call for extemporaneous remarks. From time to time, you will be called on to introduce a speaker. And in everyone's life there comes a time when a formal, prepared speech is the best way to get the job done.

The Formal Speech

The formal speech is used when you need to communicate in a formal, structured situation. Not everyone likes to speak in public. But public speaking is important in many jobs and situations. Business executives are asked to speak to community groups. Many groups work on joint projects with other groups. The ability to share ideas in a public speech helps accomplish group goals.

Think about occasions when it is important to be able to deliver a formal speech. Here are a few examples: a report to your class, a committee report to your FFA chapter, a presentation to the county government, a sales report to your employer and board of directors.

An FFA member delivering a speech.

Types of Formal Speeches. In planning a speech, keep in mind the different purposes of prepared speeches. There are five basic types of formal speeches. The first three generally are used most frequently in meetings.

The speech to *inform* or *instruct* is used to speak about a specific topic. It could be a report of activities or an explanation of, or directions for carrying out, an activity. A local agricultural organization might ask you to speak on the activities of your FFA chapter.

You try to *persuade* or *convince* in remarks that seek people's agreement to a position, a decision, or a course of action. Political speeches are designed to persuade: ''I should be elected because I know farm problems firsthand.''

Action seeking tries to promote group action. Detailed reasons should be given for taking the action. The park is needed because many children are playing in the street and it would make an excellent BOAC project.

Challenge speeches often are used in a program situation at a meeting or banquet. The speaker tries to inspire the audience to improve itself or a situation. The state secretary of agriculture might challenge FFA members to help people in hungry nations learn to feed themselves.

A speech to *entertain* may be used in combination with other types of speeches. It helps change the pace at a meeting where many serious speeches are being given. Few people deliver purely entertaining speeches. But any speech can be enlivened with tasteful jokes or funny stories.

A speech does not have to be just one type. For example, challenge speeches will contain information, seek action, and may contain entertaining sections.

Preparing the Speech

An effective speech is one that is well prepared. Preparation ensures quality and helps hold down nervousness. The first step is to analyze the purpose of the speech and the nature of the audience. If you are keynote speaker at a convention, you will speak more generally and inspirationally than if you are to present an informational report. Your approach would be different in speaking to a group of business executives than before a government agency.

If a topic has not been assigned, pick one that fits the purpose of the meeting or group. "Technology in Agricultural Industry" would be a good topic for an FFA speech. Ask the person who invites you to speak about the topic, speaking conditions, and the length of the speech. Will you be in an auditorium or classroom? Will there be a podium? What is the purpose of the meeting? What topics interest the group?

Often, you will receive precise instructions. But you must make sure that you know all possible information about the event at which you are speaking.

Once you know the speaking conditions, pick a general topic on which you are qualified or can become prepared to speak. Then gather your reference materials from books or other sources. Be sure to check your facts for accuracy. Select appropriate illustrations and stories to help bring your speech alive. Would the use of visual materials, such as pictures, charts, or maps, be appropriate?

Suppose you have been asked to speak on backpacking and camping: "It is important in hiking to be able to find your way. Someone asked Daniel Boone if he ever got lost. 'No, but I've been mighty confused for a couple of days,' Boone replied. We don't need to be lost or confused if we learn how to use a map and compass."

Keep your topic to manageable size. If asked to speak on the history of agriculture, talk about important developments in soil conservation or farm machinery. You don't need to start with the first herders and sowers.

Writing the Speech

Organization of the speech is a key to success. Experienced speakers may outline short talks on paper or mentally. But beginning speakers should write out all speeches in detail. They may then use an outline to speak from. It helps to approach the podium with a few notes to serve as reminders or clues. Also, develop an outline before writing the speech. It helps to make sure the speech is organized and that one part follows another. A sample outline follows:

Topic: Technology in Agricultural Industry

I. Introduction: Influence of technology on increased efficiency of food production.
   A. Productivity of American farmers.
      1. End of Civil War, one farmer produced food for one family—four to six people.
      2. By 1920, one farmer produced food for ten people.
      3. By 1970, one farmer produced food for fifty people.

II. Body.
   A. Mechanization leads to increased productivity.
      1. Change from animal and human power to machine power.
      2. Development of new and more efficient equipment: seeders, harvesters, etc.
   B. Research plays a part.
      1. Colleges of agriculture, federal government, private industry.
      2. Breeding plants resistant to disease.
      3. Improved production of animals.
   C. Technology.
      1. Yields have gone up with increased use of commercial fertilizers.
      2. Efficient irrigation equipment available to meet water needs.

III. Conclusion: Education a key.
   A. Meeting the challenges of the future depends on education.
   B. Study of plants, animals, mechanics, leadership.
      1. School.
      2. FFA.
      3. Supervised occupational experience.

Introduction. This creates interest in the topic and tells your audience what you are going to tell them. Relate the topic to the audience and its experience: "As businesspeople, you know the importance of efficiency in production. Tonight, I will speak on the influence that agricultural technology has had on the increased efficiency of food production in this country."

The Body. This contains your message: the facts, information, and supporting material needed to clarify your position on the issue being presented. Organize the body of the speech in a logical sequence with the key points highlighted. Your speech to businesspeople could describe, in order, (1) how mechanization has increased production, (2) advances in crop and livestock production, and (3) how technology has increased the efficiency of food production in America.

Conclusion. This permits you to review the key points of the speech. You may give a summary, mention the themes of the speech, and quickly review your facts. End with a strong concluding statement: "So you see, the development of farm and related technology was important in making our system for the production of food the most efficient in the world."

Before the Speech

After the speech is written, there are two more important activities before it is actually delivered: review and practice.

Reviewing the Speech. Analyze the speech carefully before delivering it. You may also ask a friend or teacher to read it. This is to ensure that it is accurate, interesting, and organized. Only you can do this job. Do the parts flow together? Do your facts support your main points? Are grammar and sentence structure correct? Does the introduction introduce the topic

—4—

Judges Score Sheet

Rev. 1965
Form 651

NATIONAL PUBLIC SPEAKING CONTEST — FUTURE FARMERS OF AMERICA

Part I. For Scoring Content and Composition:

Items to be scored	Points Allowed	Points awarded contestants												
		1	2	3	4	5	6	7	8	9	10	11	12	13
1. Content of the manuscript	200													
2. Composition of manuscript	100													
Score on written production	300													

Part II. For Scoring Delivery of the Production:

Items to be scored	Points Allowed	Points awarded contestants												
		1	2	3	4	5	6	7	8	9	10	11	12	13
1. Voice	100													
2. Stage presence	100													
3. Power of expression	200													
4. Response to questions	200													
5. General effect	100													
Score on delivery	700													

Part III. For Computing the Results of the Contest:

Items to be scored	Points Allowed	Points awarded contestants												
		1	2	3	4	5	6	7	8	9	10	11	12	13
1. Score on written production	300													
2. Score on delivery	700													
TOTALS	1000													
*Less overtime deductions, for each minute or major fraction thereof	20 Points													
*Less undertime deductions, for each minute or major fraction thereof	20 Points													
GRAND TOTALS														
Numerical—or Final placing of Contestants														

*From the timekeeper's record.
For explanation of score sheet items, see Section IX, page 3.

A judges score sheet for an FFA public speaking contest.

and will it make the audience want to hear the rest of the speech?

Have you used statistics sparingly, accurately, and only to illustrate points? Do quotations support your themes or are they scattered into the speech without much apparent reason? Does the title sound interesting? A good title helps whet the appetite of your audience.

Then, rewrite and revise. Beef up weak areas. Fill in any gaps. Explain things that seem vague and unclear. Cut material that is not related to the topic. Move ideas around so they flow together in an organized whole. Speeches take a lot of hard work. It will help if you look up and read several speeches. Your library may have a collection of speeches.

Practicing the Speech. "Practice makes perfect," according to an old saying. You need not be perfect, but practice will help you deliver your speech well. If the speech is short, review the outline several times. For longer speeches, you should rehearse several times. If you are speaking in an FFA speaking contest, you must follow the text that judges watch. See the score sheet for points that judges of FFA contests use to evaluate speeches.

Otherwise, you may want to try to practice speaking from the outline. Refer to the written speech when you are unsure. Work hard on the spots you have trouble remembering. Speak out loud to your mirror and/or your family. Except for FFA contests, you need not remember every word, but should have the main ideas fixed in your mind.

Use a tape recorder to make sure you are speaking loudly and clearly enough and that you have animation in your voice. Also, rehearse your gestures. Points can be emphasized with your hands and facial expressions.

Self-Confidence. A little tension is to be expected before a speech. But if you have done adequate preparation, you know the speech well and that it is good. While waiting to speak, watch what others are doing. Check out arrangements one more time. How do you get to the podium? If you are neat and well groomed and feel your best, you will do a good job. Take a few deep breaths on the way to the podium to ease any remaining tension.

Delivering the Speech

Stand erect and look directly at the audience, not over their heads. Seek and maintain eye contact as you look at different parts of the room. Don't lean on the podium. Slightly bent knees reduce tension and fatigue. If you stand squarely on both feet, you will feel solid and probably give a more solid speech.

Vary your volume and tone as you speak. Gestures emphasize points or indicate an illustration. Point to the spot on the map or the part of the diagram you are discussing. If you have planned and rehearsed the gestures, they will feel and appear natural and not detract from the speech. Adjust your facial expression to the tone of the talk. Be pleasant, but don't smile at serious points or frown when you're telling a joke. Let the audience laugh at your jokes.

Voice, expression, gestures, and body position communicate sincerity and conviction to the audience. They make up the nonverbal parts of your speech. You can tick off points on your fingers. A clenched fist can show determination. A sweep of the hand indicates the breadth of what you are talking about.

After the Speech

Honestly evaluate how you did. Make a list of the strong and weak points of the speech

Note the points about delivering a speech.

and your delivery. How can you improve the weak points? Analyze comments from your listeners. Ask friends who listened to the speech to honestly evaluate your performance.

Could you be heard in all parts of the room? Were your main points clear? Did your facts support the main points? Were your stories and illustrations appropriate? Do you have any nervous habits that were distracting?

These postmortems are important. People who speak frequently in public often have their colleagues or friends help them with after-the-speech evaluation and improvement.

Extemporaneous Speaking

You are at an FFA banquet. You have just finished the main course when the president whispers to you: "The guest speaker is delayed. Could you speak a few minutes after dessert?" You ask, "What will I talk about?" The president says, "You worked on the program of activities. Just give a brief outline of that."

Since you do not have time to prepare your remarks, you will have to speak *extemporaneously*. This happens fairly often. The topics of extemporaneous speeches generally relate to the speaker's area of interest—FFA activities in your case. You are selected because of your knowledge of the subject. You don't need to make a long speech, but your thoughts must be organized to get your points across to the audience. In the few minutes before dessert and coffee, you can organize an introduction, the body of remarks, and a summary. Prepare a brief outline that covers your main points:

I. Intro: Busy year ahead; activities in all areas; shows we have an active chapter.
II. Body: This is kickoff event.
  A. Community service.
    1. Park maintenance project.
    2. Tree planting project.
    3. Voter registration first time this year.
  B. Full program of leadership activities.
  C. Other related activities.
    1. State convention participation.
    2. Projects.
      a. Equipment maintenance.
      b. Seed experiments.
III. Summary: Active all FFA areas; more projects than last year. Thanks for your attention.

The outline helps you make sure you speak in an organized fashion and cover all major points. A few key words remind you of the areas you want to address. In speaking extemporaneously, follow the same principles as in any public speech.

Answering Questions

After any public remarks, there may be questions the audience wants to ask. Answering questions is an important scoring item in FFA public speaking contests. It can determine the winner of the speech.

Questions may seek additional information or clarification of points made in the speech. Your advance work must prepare you for this part of public speaking. Your research will turn up information that may be interesting and related, but not important enough to include in the speech. File this information away in your mind. Also, try to think through in advance what questions might be asked and work out and rehearse your answers.

Often, you are in charge during the question period. Stand erect and try to watch all parts of the room for hands that are raised. Point to the person and say the name if you know it. If there is likely to be confusion, be specific about who you mean: "The gentleman in the blue suit in the middle of the third row."

Introducing a Speaker

An active group leader often is asked to introduce speakers. These introductions usually are extemporaneous, but should be planned and prepared. Introductions should be brief, concise, and accurate. The speaker arrives, and is greeted at the door: "Hello, Mr. Booth, I'm Donald Roberts, chapter president. I have the honor of introducing you this evening." Visit with the speaker to learn enough details to make the introduction. Often, these details can be obtained from a biography, if one is provided. Your introduction sets the stage for the speaker. Be enthusiastic and sincere. Organize the introduction as follows:

- A brief review of the speaker's background and qualifications. Why is this person chosen to speak?
- Relate the speech to the group.
- Give the exact title of the speech.
- Give the name of the speaker and his or her title.
- Begin the applause and take your seat after the speaker comes to the podium.

Here is a short outline of an introduction:

I. We are pleased to have our local banker here.
   A. President of First National Bank, past 6 years.
   B. Before that, in loan department 8 years.
   C. Member, various bankers' groups, Chamber of Commerce, and local civic club (give name).

Representative Ronald V. Dellums speaking at a public gathering.

D. Spoke to us last year on finance.

II. Our current unit deals with securing loans.

A. We have discussed sources of credit and interest rates.

B. Also discussed completing loan application forms.

III. Fortunate to have the opportunity to hear about "Securing a Loan."

IV. We are pleased to have as our speaker, the President of the First National Bank, Mr. Thomas A. Booth.

Communication for Living: A Review

All forms of group communication are important as you work with others. The American democratic system depends on skill in group communications. Government officials often share information and announce decisions in public speeches or in press conferences. The President may announce a change in farm policy at a meeting of an agricultural organization.

You can't leave things up to others. Everyone needs to develop the ability to communicate with groups. This is particularly important in business and with community organizations. Clear communications make businesses more efficient and the community a better place to live. Developing group communications skills is one more way to assume responsibility for your own life and for the benefit of groups you have joined.

THINKING IT THROUGH

1. Why is it important to be able to communicate with groups of people?
2. Why do people sometimes misunderstand what the speaker intends to say?
3. What is the difference between a discussion and a conversation?
4. What are the ways of being an effective discussion participant?
5. What are the ways of being an effective discussion leader?
6. What are the major purposes of speeches?
7. What should be done to prepare for a speech?
8. How should a speech be presented?
9. What does extemporaneous speaking mean?
10. How is a speaker introduced?

CHAPTER 9

PARLIAMENTARY PROCEDURE

Holly was frustrated and embarrassed. She was frustrated because the meeting of the Garden Club was out of order and embarrassed because she didn't know how to get it back in order. People weren't sticking to the subject, a project to supply flowers to patients in local hospitals. Grace was making a long speech on the need to organize next year's flower garden contest. And Steve had just finished talking about people who were habitually late with their dues. The project had been on the agenda for three meetings, with no decision.

This group lacked any procedures or guidelines for conducting its business. As in many organizations, the members thought parliamentary procedure would limit freedom of speech. However, *parliamentary procedure* is one set of procedures that does provide a chance for everyone to speak. At the same time, it provides a way to get a group's business done. These procedures protect everyone's freedom, not just the loudest and most persistent.

Most Americans believe that each person has a right to speak and take part in meetings. This is the soul of democracy. But the group also has a right to get its business done. For effective operation, groups generally follow the basic guidelines of parliamentary procedure. The ancient Greeks had rules to govern their meetings. So did the British Parliament centuries later. Many of these rules were adopted by the U.S. Congress and state legislatures.

Skill in using parliamentary procedure is another tool needed by members and leaders who want to take part in and guide group meetings. The basic guide for conducting meetings is *Roberts' Rules of Order*, prepared by General Henry M. Roberts in 1876. He simplified the rules of procedure used by the U.S. House of Representatives so they could be used by any organization. The material in this chapter is based on his work.

CHAPTER GOALS

In this chapter your goals are:

- To explain the purposes and functioning of parliamentary procedure
- To classify a list of motions as to type and use
- To describe how to obtain the floor at a meeting

- To introduce a motion during a meeting
- To demonstrate the correct use of parliamentary procedure in a mock meeting

Purposes of Parliamentary Procedure

The rules of parliamentary procedure were developed over the centuries to meet specific situations in group debate. The rules make meetings more democratic. They try to balance the need to get the majority's business done against the need to protect the rights of the minority. Neither set of rights is absolute. A person who opposes the majority has the right to be heard, but not to block action. The majority has the right to reach a decision, but not at the price of silencing the minority. Parliamentary procedure has evolved to serve the following basic purposes in conducting meetings:

- To make sure one thing is considered at a time
- To ensure courtesy to everyone
- To ensure the will of the majority
- To protect the rights of the minority
- To ensure that everyone who wants to speak will have the opportunity

In the hands of a skillful presiding officer, these rules of procedure can ensure that a proper balance is struck between the rights of the majority and minority, between adequate debate and too much debate, and between getting things done and ensuring that everyone is heard.

Learning parliamentary procedure has practical applications. Many groups use it to get their business done. While many businesses do not use these rules in meetings, practice in parliamentary procedure helps you learn to stick to the subject, respect the rights of others, and move to a decision. Business meetings do have a structure and rules, even if the guidelines are not formal.

Essential Features of Parliamentary Procedure

To the uninformed, parliamentary procedure may seem arbitrary. A person who does not understand the rules and leaps up to speak while another is speaking may be angry when the presiding officer enforces proper procedure by a rap of the gavel and an order to be seated.

However, parliamentary procedure is, in fact, fair. Its primary characteristic is impartiality. By having rules that apply equally to everyone, the rights of everyone are ensured. During a discussion, each person has an equal right to be heard.

Secondly, parliamentary procedure establishes an order in which to consider the group's business. For example, a motion to amend, or change, another motion must be disposed of before the motion to which it applies. The FFA chapter must vote on the amendment concerning whether the tractor to be purchased is to be new or used before voting to buy the tractor.

Thirdly, parliamentary procedure makes sure that only one question at a time is considered. If the motion is on buying a tractor, then the group may not discuss the year's supervised occupational experience program.

Finally, there is a procedure for moving to a vote. If a member persists in talking about the virtues of one make of tractor over the other when the rest of the group is ready to act, a motion can be made to call for the previous question. This motion requires a two-thirds vote to pass. Thus, any sizable majority can continue the debate until all of its concerns are expressed.

Parliamentary procedures provide order at meetings with only one person speaking at a time.

How to Use Parliamentary Procedure

Every meeting should have an *agenda*, which is an organized order in which business is conducted. A typical agenda might include the following:

- Call to order and/or opening ceremony
- Roll call of members and announcement of attendance
- Reports of secretary and treasurer
- Committee reports
- Unfinished business
- New business
- Program
- Adjournment and/or closing ceremony
- Refreshments and/or recreation

Most of the group's business will be done under the agenda items *committee reports*, *unfinished business*, and *new business*. A number of committee reports may be recommendations for action. These recommendations can be debated and voted on. Unfinished business is action that was left over from previous meetings: "We did not have time last meeting to complete the discussion of our crop-production project. I bring it up now under old business." New business gives group members a chance to raise issues that concern them. An issue may be referred to a committee for study. For example, if a member proposes a raise in dues, the president could appoint a committee to report on the issue at the next meeting.

Conducting a Meeting

Call to Order. The presiding officer calls the meeting to order at the designated time and determines if a quorum is present. In the FFA, the call to order is followed by the opening ceremony, which is found in the FFA *Official Manual*. Other groups may open their meetings with the national anthem or The Pledge of Allegiance to the Flag.

A *quorum* is the number of people who must be present to do business. It usually is defined in the constitution, or bylaws of a group. Usually, it is a majority of the membership. If enough members are not present and cannot be found, the meeting must be recessed or adjourned.

Presiding at a Meeting. The presiding officer, or *chair*, is responsible for keeping order. The agenda for the meeting is followed by calling out the next order of business: "It is now time for unfinished business." This item may have several subheads. The chair says, "The first item under unfinished business is finding a new meeting room."

During debate, the chair assigns the floor, which means choosing who speaks in what order. If this job is not done well, members may become angry and try to disrupt the meeting. They feel the chair is being arbitrary. The chair should ensure that persons both for and against the motion are given an opportunity to speak.

The chair also calls for, and announces the results of, voting. It is important not to call for a vote until every member has spoken who wants to. The chair says, "The vote is on the motion to visit the County Council as part of our citizenship project."

Correct use of the gavel is important. It is the symbol of authority that calls attention to business. *Two* taps call a meeting to order. *One* tap announces a vote or a decision of the chair. *Three* taps signal members to stand and *one* to be seated. Several sharp taps are used to restore order in a meeting.

Obtaining the Floor. This is gaining the right to speak. To obtain the floor, you stand when another member has finished speaking and say "Mr. President" or "Madam President." The chair responds by calling the name of the person who is to speak. The chair must choose if more than one person seeks recognition at the same time.

Introducing a Motion. New business can be introduced in the form of a motion. A *main motion* introduces the business—to buy a tractor, for example. *Subsidiary motions* can amend the main motion, delay its consideration, or refer it to a committee for further study.

To make a motion, obtain the floor and state it clearly: "I move that the dues of the organization be raised 50 cents." Most motions require a *second* to show that another person thinks the idea is worth considering. You don't need to be recognized to second a motion. Call out, "I second the motion."

The chair then restates the motion to give the person who made it a chance to check its accuracy. This also ensures that all members understand the motion. The floor is then open for further action. This may be the debate and vote to accept or reject the motion. However, some motions are not debatable, and the chair calls for an immediate vote on them.

Debating a Motion. The debate must relate to the motion on the floor. Speakers who wander off the subject should be ruled out of order by the chair. If an FFA chapter were debating a motion to go on a field trip, it would be unrelated and out of order to discuss the program for the banquet. This rule of germaneness ensures that the motion itself is acted on and that unrelated questions are not used to delay or obstruct the business of the organization.

All members can speak on the motion. The presenter usually speaks first. While the chair should ensure that all sides are heard, debates should not be allowed to ramble on after everyone has had a say.

At the end of debate, before voting starts, the chair should restate the motion. This ensures that all members know what they are voting on. Silence on the part of the members at this point indicates they are ready to vote. Once the voting starts, a member may not continue debate.

Voting on a Motion. Most motions can be disposed of with a voice vote. The chair says, "Those in favor of the motion say aye." After the aye vote, the chair says, "Those opposed say no." Then the result is announced.

However, if the debate indicates a close vote or the voice vote is too close to count, the chair can move to other types of voting. One type involves a show of hands or a standing vote. The chair states, "Those in favor of the motion raise your right hand [or stand]." "Those opposed please raise your right hand [or stand]." If the outcome is obvious, the chair an-

One method of voting is by raising hands.

nounces the result. If the vote seems close, the chair should take a count of hands or of those standing.

The roll-call vote and ballot are the most precise methods of voting. They may be called for by a group's bylaws or when a majority of members vote to use either form.

In roll-call votes, each member's name is called in alphabetical order. The member responds aye or no, or abstains. The roll call provides a record of how members voted. In the case of a ballot, the secrecy of each member's vote is ensured. The chair assigns members to distribute, collect, and count the ballots.

A *majority vote* (one more than half the members present) is required in an election or for passage of most motions. A tie vote is lost. The chair may vote to make or break a tie. A member may change her or his vote only up until the time that the vote is announced by the chair.

Nomination of Officers. *Nominations* name a person as a candidate for office. They may be made from the floor, by a committee, or by a nominating ballot. Nominations are like other motions, but do not require a second. *Nominating committees* may suggest one or more names for each office. All candidates should be asked if they are willing to serve. To make a nomination from the floor, rise and say, "I nominate Mike Cranmer for president." Nominations may be closed by the chair if no additional nominations are heard. If a member objects to the close of nominations, it must be put to a vote. The motion to close nominations requires a second and a two-thirds vote for passage.

Types of Motions

Motions are the vehicles for group decisions. They provide the course of action that is chosen or rejected. There are different types of motions. Table 9-1 outlines the procedures and requirements for each type. For more detail, consult *Roberts' Rules of Order*.

The Main Motion

The main motion brings a suggested action or activity before the group. For example, "I move that the FFA chapter buy a tractor." To ensure order, only one main motion can be debated at a time. Some action must be taken before another main motion can be considered. The main motion is the lowest ranking of all motions in terms of precedence. This is so because other motions affect the main motion and must be dealt with first. The main motion is debatable, can be amended, and can have subsidiary motions applied to it.

Subsidiary Motions

Subsidiary motions change or dispose of the main motion and take precedence over it. They are outranked by privileged and incidental motions. Following is a discussion of the more common subsidiary motions in order of precedence.

Lay a Motion on the Table. This delays action until the motion is removed from the table. It gives the group time to gather information or to consider more urgent business. The question must be removed from the table later in the same meeting or by the end of the next regularly scheduled meeting. It requires a second, may not be debated or amended, and requires a majority vote to pass.

Previous Question. This motion, if passed, ends debate and leads to an immediate vote on the motion before the group. It requires a second and cannot be debated or amended. Since it limits debate, it requires a two-thirds vote for passage. If the motion calling for a vote on the previous question passes, the chair immediately calls for a vote on the previous question itself. The chair should explain this procedure carefully so that members aren't confused as to what they are voting on. The previous question refers to the last motion on the floor unless it is stated to include all pending business. If a main motion and two amendments are on the floor, all pending business would include both amendments and the main motion and would involve three votes.

Limit or Extend Debate. This motion sets or adjusts the amount of time allowed for debate. It requires a second, cannot be debated, but can be amended. A two-thirds vote is required for passage.

Postpone to a Certain Time. This motion, if passed by a majority, delays action until the time set in the motion. It requires a second, is debatable, and can be amended.

Refer a Motion to a Committee. More information may be needed before a final vote on the main motion. In the case of the sample main motion, members may want to know why the group needs a tractor, how much it costs, the difference between buying a new or used tractor, what additional equipment is needed. The motion would be referred to committee for further study. The motion to refer to a committee, which is debatable and amendable, requires a second and a majority vote for passage.

The committee might be asked to re-

TABLE 9-1. Parliamentary Procedure[1]

MOTION	IN ORDER WHEN ANOTHER HAS THE FLOOR	REQUIRES A SECOND	DEBATABLE	AMENDABLE	VOTE REQUIRED	CAN BE RECONSIDERED
MAIN MOTION	No	Yes	Yes	Yes	Majority	Yes
PRIVILEGED MOTIONS (In precedence)						
Fix time at which to adjourn	No	Yes	No	Yes	Majority	Yes
Adjourn	No	Yes	No	No	Majority	No
Take a recess	No	Yes	No	Yes	Majority	No
Question of privilege	Yes	No	No	No	None	No
Call for orders of the day	Yes	No	No	No	None	No
INCIDENTAL MOTIONS (No order of precedence)						
Appeal decision of chair	Yes	Yes	Yes[2]	No	Majority	Yes
Close nominations	No	Yes	No	Yes	2/3	No
Division of the assembly	Yes	No	No	No	None	No
Division of a question	No	Yes	No	Yes	Majority	No
Object to consideration	Yes	No	No	No	2/3	Yes
Parliamentary inquiry	Yes	No	No	No	None	
Point of order	Yes	No	No	No	None	No
Request for information	Yes	No	No	No	None	No
Suspend the rules	No	Yes	No	No	2/3	No
Withdraw a motion	No	No	No	No	Majority	Yes
SUBSIDIARY MOTIONS (In precedence)						
Lay on the table	No	Yes	No	No	Majority	No
Previous question	No	Yes	No	No	2/3	Yes
Limit or extend debate	No	Yes	No	Yes	2/3	Yes
Postpone to a certain time	No	Yes	Yes	Yes	Majority	Yes
Refer to a committee	No	Yes	Yes	Yes	Majority	Yes
Amend	No	Yes	Yes[2]	Yes	Majority	Yes
Postpone indefinitely	No	Yes	Yes	No	Majority	Yes
OTHER MOTIONS						
Ratify	No	Yes	Yes	Yes	Majority	Yes
Reconsider	Yes	Yes	Yes[2]	No	Majority	No
Rescind	No	Yes	Yes	Yes	2/3	Yes
Take from the table	No	Yes	No	No	Majority	No
To make nominations	No	No	Yes	No	Majority	

[1]Adapted from *Roberts' Rules of Order.*

[2]Yes, when applied to a debatable motion.

state the motion to recommend action: "I move that the question be referred to a committee of four appointed by the chair to investigate the need for, and cost of, a tractor and to report back to our next meeting with a recommendation."

Amend a Motion. An amendment is used to change the main motion. The change must be related to the main motion. If the group is debating a motion to buy a tractor, it is out of order to offer an amendment fixing the date for the annual banquet, which is another main motion. Amendments can be presented at any time during the discussion and must be seconded. Many amendments can be made to a main motion. However there can only be one amendment to an amendment pending at a time. There could be an amendment to add the words *and cutter bar* to the motion to buy the tractor (primary amendment) and then a secondary amendment to insert the word *used* before *cutter bar*. The vote must be taken first on the secondary amendment. Then another secondary amendment can be made.

Using the motion "to buy a tractor and plow" as an example, amendments may take the form of:

- Adding: "I move to amend the motion that we buy a tractor and plow by adding the words *and cutter bar*."
- Inserting: "I move to amend the motion by inserting the word *used* before *tractor and plow*."
- Striking out: "I move to amend the motion by striking out the words *and plow*."

Amending a motion.

- Striking out and inserting: "I move to amend the motion by striking out the word *plow* and inserting the words *cutter bar*."

Substituting is another form of an amendment. It is in order only when it is desired to change an entire resolution or an entire paragraph of a section or a resolution. The last amendment pending is voted on first. Then the next amendment, and then the motion as modified by the amendments that passed. Assume the motion to buy the tractor was followed by amendments to buy a cutter bar and to make it a used tractor. The order of voting would be: (1) whether the tractor should be used, (2) whether to buy the cutter bar, (3) the main motion as amended by any of the other amendments that passed. The floor should be opened for discussion after each amendment is voted on.

Incidental Motions

These motions clarify a question from which they arise. They can concern the procedure used, the room conditions, or information related to the discussion. They take precedence over the question before the group because they affect how the question is to be handled. They must be acted on before the question, but have no relative rank among themselves. Following is a discussion of some of the more common incidental motions.

Appeal from a Decision of the Chair. An appeal from the decision of the chair seeks to reverse a decision by the presiding officer. This motion requires a second and a majority vote for passage. A tie vote defeats the motion and upholds the chair's decision. For example: "The chair has ruled that the motion to add 5 acres [2 hectares] to the experimental plot is out of order because it does not relate to the motion to buy a tractor. Those in favor of sustaining the decision of the chair say aye, those opposed, no."

Division of the Assembly. This motion is used to determine the accuracy of a vote. It calls for a recount, usually of a voice vote. The new vote probably would be a show of hands or a standing vote. The motion does not require a second and is not debatable or amendable. To make the motion, state, "I call for a division."

Parliamentary Inquiry. You seek information about procedure by saying, "I rise for a paliamentary inquiry." The chair answers your question or gets the information you need. No second is needed.

Point of Order. If you think the rules are not being enforced by the chair, stand and say, "I rise to a point of order." A point of order is usually decided by the chair. It is not debatable or amendable and does not require a second.

Request for Information. This motion can be directed to the chair or another person. It seeks information about the matter under discussion.

Privileged Motions

Privileged motions deal with an emergency situation or urgent matters. They have precedence over all other types of motions and are not debatable. They rank among themselves as follows:

- *To fix the time at which to adjourn* sets a definite time to end the meeting
- *To adjourn* closes the meeting
- *To take a recess* provides for a short break in the meeting

Raising a question of privilege.

- *To raise a question of privilege* is used to obtain immediate action to correct a problem situation
- *A call for the orders of the day* is used to keep the organization on the regular order of business

Using Parliamentary Procedure: A Review

Parliamentary procedure is followed by many groups and local, state, and national legislative bodies. It may be used also by the boards of directors of some businesses, but not as much in the day-to-day operations of business. It often is used in meetings where large groups of people are involved or where opinion is likely to be sharply divided. It is less likely to be used in informal meetings.

Even though you don't use parliamentary procedure every day, learning and practicing it can be important. It gives firsthand experience in respecting the rights of others, doing business in an organized and orderly fashion, and acting democratically. FFA chapters often hold practice sessions to give each member a chance to practice proper parliamentary procedure as a member and as the presiding officer. A working knowledge of parliamentary procedure prepares you to take part in meetings.

THINKING IT THROUGH

1. What are the purposes of parliamentary procedure?
2. What are the essential features of parliamentary procedure?
3. What is the usual order of business for a meeting?
4. How can a member obtain the floor?
5. Why is a working knowledge of parliamentary procedure important to the chair?
6. How should a motion be introduced?
7. How can the gavel be used?
8. What should be considered in debate on a motion?
9. What are the ways of conducting a vote?
10. How are nominations handled?
11. What are the types of motions?
12. Why do some motions have precedence?

UNIT III

DEVELOPING LEADERSHIP SKILLS

COMPETENCIES	PRODUCTION AGRICULTURE Farm manager	AGRICULTURAL SUPPLIES/SERVICES Agricultural chemical salesperson	AGRICULTURAL MECHANICS Agricultural mechanic	AGRICULTURAL PRODUCTS, PROCESSING, AND MARKETING Dairy plant worker	HORTICULTURE Garden center manager	FORESTRY Forest service supervisor	RENEWABLE NATURAL RESOURCES Recreation farm manager
Demonstrate the ability to work cooperatively	Very Important	Very Important	Very Important	Very Important	Very Important	Very Important	Very Important
Exhibit the ability to work as a member of a team	Very Important	Very Important	Very Important	Very Important	Very Important	Very Important	Very Important
Develop a knowledge of the organization of which you are a member	Very Important	Very Important	Important	Important	Very Important	Important	Very Important
Describe how personal and professional needs can be met within an organization	Important	Important	Important	Important	Important	Important	Important
Show a good attitude toward the group	Very Important	Very Important	Very Important	Important	Very Important	Very Important	Very Important
Demonstrate the ability to communicate effectively	Very Important	Very Important	Important	Important	Very Important	Very Important	Very Important
Exhibit the ability to participate in a meeting	Very Important	Very Important	Important	Important	Very Important	Very Important	Very Important
Supervise the activities of others	Very Important	Important	Important	Important	Very Important	Very Important	Very Important

Very Important

Important

Not Important

COMPETENCIES	PRODUCTION AGRICULTURE Farm manager	AGRICULTURAL SUPPLIES/SERVICES Agricultural chemical salesperson	AGRICULTURAL MECHANICS Agricultural mechanic	AGRICULTURAL PRODUCTS, PROCESSING, AND MARKETING Dairy plant worker	HORTICULTURE Garden center manager	FORESTRY Forest service supervisor	RENEWABLE NATURAL RESOURCES Recreation farm manager
Develop the ability to assume responsibility							
Show that you are able to benefit from supervision							

Personal growth and leadership development do not occur in a vacuum. They take place in situations that involve other people. Much growth and some leadership development can occur in your daily interactions with friends and family members. But developing your abilities fully requires something more. You need a place where you can try out the skills and abilities that you study and work on in private. To communicate effectively, you need to face other people.

You must relate to and communicate with people all the time. This gives you a great deal of practice at communications skills. But you can do a little extra to acquire and sharpen the abilities that you will need all your life. That little extra can be work with an organization.

You probably already belong to organizations such as the Scouts, 4-H, organized sports teams, music or drama clubs, FFA, FHA, or a discussion group. You probably joined one of these organizations because you enjoyed its activities. Or you may have liked the people and wanted to be friends with them. Often, your friendships and interests will coincide.

But maybe you're not making the most of your opportunities. Have you considered the possibilities your organizations provide for personal growth and leadership development? Do you try to have input into the decisions of the organization? Do you take an active part in its activities? Do you do your share of the work? Have you sought a position of formal or informal leadership?

Organizations can be laboratories for developing personal and leadership abilities. Maybe you need a group that fits your personal needs more closely—one that combines leadership development with your vocational interests. If this is the case, you'll want to look around to see if such an organization exists.

Such a laboratory would be ideal preparation for finding and doing a competent job. Many of the skills you learn in an organization are necessary for keeping a job and advancing in it. The matrix on the preceding pages is based on this knowledge and these skills.

People take part in meetings all their lives. Many feel uncomfortable because they passed up opportunities when they were younger to develop their leadership abilities. Now they cannot get the most out of their work or participation in the PTA, Kiwanis Club, religious boards, farm organizations, or other groups.

You have the opportunity now to learn what it really means to become a member of an organization. What part will you play in this organization? What are your duties and responsibilities? What does leadership mean to you?

Your overall goal in this unit is to learn how to evaluate your needs and to match them with that of an organization. This includes understanding what your responsibilities will be at different levels of involvement, as a member and a leader. In working toward this goal, you will be laying the groundwork for a full and satisfying life as an employee of a business organization and a member of organizations which serve both your personal needs and those of your community.

CHAPTER 10

EVALUATING ORGANIZATIONS

Chris joined the Cedarville Farm Co-op because it provided things he needed to raise corn on his farm. The co-op provided high-quality products and services at reasonable prices. Chris always found the co-op ready to stand behind the seed corn, fertilizer, and equipment it sold. There also was a profit-sharing plan for the members of the co-op, who, like Chris, were farm operators.

The farm co-op was an organization whose primary purpose was to serve its members, who also were its customers. Businesses are organizations. Some are membership organizations. Most are organizations that serve their owners with profits, their employees with income and satisfying work, and their customers with goods and services. Like all organizations, businesses are not successful if they do not serve their members or customers.

Much of your life will probably be spent with a business organization. How well you choose this organization will determine your income and your satisfaction at work, as well as your opportunities for personal growth. You can learn how to choose wisely by practicing now to make intelligent choices of the organizations in which you become active.

CHAPTER GOALS

In this chapter your goals are:

- To describe the basic framework of an organization
- To analyze factors that can be used to evaluate an organization
- To evaluate and plan for fund-raising activities
- To select and take part in the activities of an organization
- To assess your needs before choosing an organization

Assessing Your Needs

The first thing to do when choosing an organization is to assess your own needs. Think through your reasons for joining. If you are clear about your own goals and purposes, you can decide which organization is best for you.

Think about the purposes and objectives you can advance by joining an organization. What values and beliefs do you have that can be advanced by an organization? What can you gain by joining a particular organization? If your vocational interests are important to you, you would want to join a group that advanced those interests.

Also consider the kind of people you want to associate with. You will want people with common interests. Think through what you can learn from members of the organization and what you have to offer them. If you are an active, outdoor type, you won't want to join a discussion group.

Activities are designed to carry out the goals of an organization. Most of these will become your goals if you join. Do the activities of an organization appeal to you? Which specific activities are you looking for? Will you be able to have input into planning the activities?

How much time are you willing to spend at meetings and in organizational activities? If you join a group, you have a responsibility to be active.

Finally, determine the areas in which you want to develop your abilities. Maybe you feel uncomfortable in front of groups and need to try public speaking. Do you have leadership abilities you want to practice and develop? Think about how you want to serve your community and other people.

The basic preparation you are engaged in now will provide practical experience that will give you confidence when you look for a job, which is discussed in Chapter 13. As in all other areas of your life, the first step in joining an organization is knowing yourself. Nobody can do the job of self-evaluation for you. Your experience in any organization depends on how well you match your own needs and

The first thing to do when deciding to join an organization is to assess your own needs.

abilities with what the organization has to offer. This is true whether the organization is a ski club or the business where you want to build your career.

Relating Your Needs to an Organization

After assessing your own needs, you will want to see how they relate to the organization you want to join. What does it have to offer you? You will want to find out about its background, its purposes and activities, and the opportunities it offers for recognition and personal growth. This information may be available in written form, or it may be obtained by talking with members of the organization or with other people who deal with the organization.

Seeking information about an organization gives you practice at skills you are learning for personal growth—planning, doing your homework, asking questions, making decisions. Even if you think you know the answers already, check out the group anyway. You may find information you didn't know existed.

Background Factors

These factors include the type and scope of an organization, its accomplishments, and its current state of health.

Type and Scope. To determine type and scope, you will look into an organization's relationship to other organizations. Some school organizations, like the FFA, are related to a particular instructional program. Others are supported and sponsored by the school for all students. In both cases, the school regards them as educational.

Other organizations are community-based. A recreation association may be sponsored by a city or county recreation department. Or an organization can have private sponsorship. Some organizations may have a religious orientation, but provide community services such as recreation, education, or citizenship training.

Check to see if the organization has state and national affiliations. The wider the affiliation, the more opportunities you will find to practice your abilities and the greater will be the challenges for you to meet.

How much support does the organization receive from the school or community? A lack of support does not necessarily mean the group is unsuitable. It may be too new to have built support. Or it may not seek support. The degree of support will depend somewhat on the purposes of the organization. A community service organization needs more support than a group that exists to serve its own members.

Accomplishments. An organization's accomplishments include activities that have been successfully carried out, the kinds of recognition provided to members, and its reputation in the community. Accomplishments should be considered in relation to the age of an organization. For example, a 5-year-old organization with five major accomplishments is more successful than a 15-year-old organization with ten.

For a community group, accomplishments would include projects successfully completed in the community. If an organization exists to maintain a park, its accomplishments will depend on how safe, clean, and useful the facilities are.

In assessing member recognition, determine what kind and how much was achieved. Did it come from the community or the group itself? Was it local, state, or national? In what form was the recognition given?

Reputation will depend on the group's record. Like support, reputation depends on how long the group has existed and how active it has been in the community. It also depends on the conduct of group members. Are they courteous and helpful? Do they keep their promises?

Current Health. Is the organization you want to join active? You want to join an organization with plenty of opportunities for personal and leadership development. Are there a variety of activities and is the group financially sound enough to sustain those activities?

An organization's strength can also be measured by the activity of its members. How many are there? How does this compare with potential membership? For an FFA chapter, the potential membership is all the vocational agriculture students in the school. How many members are active? Some groups have large numbers of members, but only a few of them are active. These organizations are not strong. In addition, members of an active organization are proud, enthusiastic, and involved. They don't complain a lot.

Purposes and Activities

Find out about the purposes and activities of the organizations you want to join. In addition to specific activities, what does the organization stand for? What does it want to accomplish? What goals does it seek to advance? Do you agree with the goals and values of the group? Do they provide you with an opportunity for personal growth and leadership development?

The organization's purposes are carried out through its activities. What are those activities? How do they relate to the purposes of the group? How, where, and to what degree does a community service organization serve the community? What specific activities serve the community?

Make sure the activities fit your needs as closely as possible. You might find a group with a terrific leadership program, but with other goals that don't fit your needs. For example, a Scout troop offers leadership development, but not much in the way of advancing an interest in vocational agriculture.

Recognition and Personal Growth

Find out if the organization provides an opportunity to develop your personal abilities such as communication, decision making, and human relations. Does it provide challenges you would not find elsewhere for participation and leadership development? Does the organization use parliamentary procedure? Ask members if they get a chance to speak and to have a voice in making group decisions. Can they become involved in activities both as members and leaders?

Everyone likes recognition, attention, and encouragement. An effective group provides recognition for members, often through an awards program. Is this true of the organization you want to join? Can all members compete for awards? Is recognition provided for people who help the group? Are there incentives and contests to promote leadership development?

Organizational Framework

The framework of an organization can be learned from its constitution and bylaws and from other documents such as organizational charts that show the relationship of titles and functions to one another. The FFA, government agencies, and many businesses have charts to give a clear picture of how the organization works.

You can obtain and examine these documents to find out how an organization is put together. For example, the U.S. Constitution establishes the way the American government is to operate. It outlines the division of responsibility and the rights of citizens. The Constitution specifies powers for the Congress, the courts, and the executive branch, and it divides power between the federal government and the states.

Like the U.S. Constitution, the constitutions and bylaws of most organizations include only the general details, giving the organizations the flexibility to meet changing conditions. Constitutions and bylaws can also be changed by a vote of the members.

In looking at the structure of your group, seek the documents that outline the sources of power and authority as well as the purposes of the group. The final power usually rests with the membership. But not always. A group may decide that the officers or directors are to retain certain powers.

Financing Organizations

You are president of an FFA chapter. The members meet to discuss the program for the coming year. Many exciting activities are discussed and the meeting ends with dozens of suggestions. The group wants to do them all. "Wait a minute," you say. "How are we going to pay for all of this?"

This is an essential question. Groups often fail because they cannot pay for things the members want.

There are three possible sources of money for an organization: donations, dues, and fund raising. Donations may come from businesses or citizens for specific purposes. However, donations are a small part of most organizations' budgets. Dues are a second source of income. But they may be kept low to encourage membership from all income levels.

Fund Raising

Low dues mean that most activities must be supported by fund raising. However, fund-raising goals should be realistic and based on a budget of actual needs.

Fund-raising activities are important to the financial health of an organization.

The Clearfield Garden Club knew how to raise money. People enjoyed its fund-raising events such as auctions. There was always a surplus in the club's bank account. But some members complained that so much effort went into raising money that little time was left to help beautify Clearfield, which related to the goals of the club.

In addition to making sure the funds are needed, some other considerations should be kept in mind.

Acceptability. Make sure the activity is acceptable to the membership and the community. The moral beliefs of the community should be respected. Bingo may be accepted in some areas, but in others people may believe all forms of gambling are wrong.

Fund-raising events should not compete with businesses or people who need to make a living. To start a nursery in a small town where there is barely enough business for the commercial nursery would be poor public relations. Auctions, fruit sales, seed sales, and personal services generally are acceptable.

Also, make sure the fund raising does not violate the constitution or policies of a sponsoring agency. In the case of an FFA chapter, for example, school policy must be checked. It may state that an event must be over at a certain time or that certain activities such as magazine sales may not be used.

Organizational Considerations. Generally, it is the best policy to relate fund raising to the purposes of the organization. This gives the fund raising a distinct quality and advertises the purposes of the organization. An FFA chapter would relate fund raising to agriculture, such as with an auction of agricultural products.

The Cedarville FFA decided to auction swine from the members' livestock production projects. This would help members and also provide the chapter a percentage of the returns. Officers and members knew hard work was needed to make the project succeed. A committee was appointed to work out the details of time, place, advertising, organization, and after-the-auction cleanup. Each committee assigned specific people to carry out specific tasks and followed up to make sure the jobs were done.

For example, the public relations committee prepared and distributed a well-written news release and made follow-up phone calls to newspapers and radio stations to make sure the release had been received and would be used. The committee also prepared flyers that were displayed in windows of businesses.

The day of the sale was planned to the last detail. Parking-lot attendants, ushers, the auctioneer, the animal handlers, cashiers, and people to deposit the money in the bank were chosen and trained. The police department was contacted to ensure adequate security and traffic control. A refreshment stand was planned to sell fruit, sandwiches, soft drinks, and cookies. People were assigned to set up the stand, to purchase and prepare the refreshments, and to sell them.

Other committees were assigned to clean up after the event and to make sure that every penny taken in at the sale was accounted for.

Organized and Efficient Meetings

Most groups hold regularly scheduled meetings, which often are specified in the constitution or bylaws. These meetings are the forum where all members can help make decisions and plan activities. They also provide an opportunity to share information. Meetings can help the group identify with and support the organization. They establish the tone and motivation for the membership.

The officers of the organization are responsible for planning and conducting efficient and interesting meetings. How to conduct meetings was discussed in Chapter 6.

Activities: Keeping Members Interested

Activities and projects are the key factors in keeping members interested. They challenge members' abilities. Activities are a major reason people join organizations. Here are some considerations in planning activities for an organization.

Relate Activities to Group Purposes

People join an organization because they are interested in its activities. You may join an FFA chapter, for example, because you are interested in agriculture and want to develop your leadership abilities. Other reasons may include a desire for friends and for more social life. But these reasons could be satisfied by joining any group. The program of activities of a group provides the specific reason for joining.

This is why the FFA provides for recognition in the area of agricultural occupations and sponsors contests that require leadership abilities. An FFA chapter should not sponsor a musical comedy or a project to repair electronic equipment. People who are interested in these kinds of activities would join a different kind of organization.

Involvement and Cooperation

The main ingredient in keeping members involved and interested is cooperation in planning activities. Some activities are carried over from one year to the next. But an effective organization will have periodic meetings to review and plan activities. In

The activities of an organization, such as an FFA chapter, are what keep members interested.

the FFA, for example, the major thrust of activities is determined when the program of activities is developed by the members.

All members should be encouraged to attend and take part in these meetings. The techniques discussed under "Encouraging Group Interaction" in Chapter 6 can be used to draw out ideas from all the members. The leaders of the organization are responsible for making sure that there is a broad enough range of activities to interest all members.

Providing a Challenge

Activities should not be planned merely for the sake of giving members something to do. Each activity should have a specific purpose and should provide a challenge for the members. Activities should involve at least some things members have not done before. This helps them learn and grow by being a member of the organization.

Often it is helpful to plan activities that bring members into contact with other groups or with people who do not belong to the group. An FFA beautification project could be planned in cooperation with the Chamber of Commerce or the local government. Members would learn how to cooperate with other groups. They also would have to use their personal and group communication skills.

Activities that serve the community or other people are a source of pride for members who take part in them. A tree-planting project can be an enduring source of satisfaction for those who took part as the trees would stand for many years as a reminder. Members would know they helped control erosion and provide natural beauty for the community.

Assessing Organizations: A Review

By assessing your own needs and measuring them against what an organization has to offer, you can choose wisely the organization in which to focus your energy and time. It is important to think broadly about your own needs, both present and future. Personal growth is an important need that can be advanced by an organization. And the challenges of leadership development provide a vehicle for that growth. By learning now to pick organizations that serve your broadest needs, you help get yourself ready to select the workplace that will suit you best.

THINKING IT THROUGH

1. What factors might be considered in evaluating an organization?
2. Why are the activities of an organization important?
3. In what ways do organizational charts or a constitution aid an organization?
4. What should be considered in selecting fund-raising activities?
5. What types of meetings and activities should be planned by members?
6. How can people select the right organizations to meet their needs?

CHAPTER 11

RESPONSIBILITIES OF MEMBERSHIP

When David became a member of the FFA chapter, he decided to be active. He volunteered for committee work and other assignments. He studied the issues that came up in FFA meetings and spoke his mind about them. David took his membership responsibilities seriously. He was elected president and selected as a delegate to the state convention. His responsible participation brought him recognition, awards, and many friends.

Donna wanted the same things as David—friends, recognition, and admiration. But she rarely attended FFA meetings, except when officers were being selected. She wanted to be an officer, but got few votes when her name was offered. Donna did not take her membership responsibilities seriously and got little back from her membership.

Membership in an organization gives a person many opportunities for recognition, prestige, and learning. But to get the rewards requires responsible participation. The organization needs such participation from members to succeed in meeting its goals. From the individual standpoint, responsible participation enables the members to get the most from membership in an organization. It is not possible to develop personally or to enjoy the services of an organization if you don't take part in the decision making and work of the organization.

Leadership is a more responsible and more committed level of membership. Leaders are still members. In general, leaders give more to an organization and members get more. However, leaders also have greater opportunity for development and advancement. One way to measure your leadership development is to consider how much you are giving to your organization in time, effort, and work.

Learning to be a responsible member and leader helps develop the ability to become a responsible employee. In the FFA, responsible participation is rewarded with awards, recognition, and advanced degrees. At work, it is rewarded with pay raises and promotions.

CHAPTER GOALS

In this chapter your goals are:

- To identify the responsibilities of a member
- To evaluate the importance of participating as a member

- To outline a procedure for electing good leaders
- To analyze the responsibilities of a leader

Responsibilities of Members

People join organizations because they share interests and goals with other members. To promote those goals requires the active participation of the membership. Responsible members take part in, and contribute to, the activities of their organization. This is the only way to understand the organization and to prepare for leadership.

Making a Contribution

You can't make a contribution if you don't attend meetings and take part in activities of your group. This also is the way you learn about the group's purposes and programs. You can help make group decisions, meet other members, and learn who the formal and informal leaders are.

> Lester wanted to be active in FFA. But he thought he didn't have time to study the issues. When the chapter debated buying a tractor, Lester didn't know what was going on. His whispered questions to his neighbor annoyed other members, and he was embarrassed when the president asked him to be quiet.

In addition to being informed, members need to make their views known at the meeting. This means issues should be studied and thought through. Homework is required if you are to speak to the point with accurate, up-to-date information.

> Dale was excited about buying a tractor for the FFA chapter. He read up on different kinds of tractors and asked farm equipment dealers about the features of different sizes and models. He learned the cost of new and used tractors. His information helped the chapter decide to buy a tractor and choose the kind that best fit its needs.

However, Dale didn't feel he needed to speak at every meeting. He was willing to be quiet if someone else presented his position. One should not speak at meetings for the sake of speaking. There are ample opportunities to be heard.

The Rights of Others. Faith was impatient. She could not wait to express her views in meetings. She often interrupted other people when they were speaking, even though the president told her to wait her turn. She laughed at people who did not speak well and made fun of people with whom she disagreed.

A responsible member respects the rights of others. This includes the right to be heard, to be treated courteously, and to disagree. Each speaker deserves the same courteous attention you expect when you are speaking.

It is relatively easy to respect the rights of someone with whom you agree. But the mature person gives a respectful hearing to someone with a different point of view. Maybe what the person says will change your mind. There may be information or situations that you had not considered or known about. And respect for others also helps ensure that everyone will get a chance to be heard. This is important in a democracy or a democratic organization.

Respecting others includes accepting and supporting the majority's decisions. Democracy depends on open debate and a full chance to disagree until the point of decision. Then everyone is expected to support the decision. If you disagree, you

An organization's goals are achieved through the active participation of its members.

can work to change it. But if everybody tried to hinder decisions they opposed, an organization could not accomplish its goals.

Committee Work. Regular meetings are only a small part of the work of an organization. Much planning and detailed work is done by committees. Responsible group members seek to work on committees. Member input on committees is important so that decisions can be made at meetings of the membership. Much of the work of decision making will have been already completed by a committee.

Because committees are small, they can deal with many details that need not occupy the full attention of the whole organization. Each committee member accepts responsibility for specific tasks. This enables committees to take the time to do in-depth studies and to report to the membership on a decision.

> Walter was a member of his FFA fund-raising committee. The committee tried to think of all the different ways to raise money. Some ideas like magazine sales were rejected because they were too time-consuming and did not have an agricultural focus. A small number of ideas were left over for further study. One was a harvest stand at the Farmer's Market in the city. Walter volunteered to investigate how to obtain a stand in the market. Other members looked into sources of fruits and vegetables and into the transportation needed for the project.
>
> The members brought their findings back to the committee. The stand was available at a reasonable price. Other members found sources of produce and farmers who were willing to share their truck space for transportation on market day. The committee prepared a report, which Walter volunteered to present to the members of the chapter. The committee recommended that the project be carried out.

This example shows that the responsibilities of a committee member are the same as those of a group member—to attend meetings, to take part in the discussion, and to work hard. Good committee members do their homework thoroughly. No gaps should be left in the information they collect to present to the membership. This requires that all information be evaluated before being placed in the report. Is the person who is willing to share the truck reliable and likely to keep promises?

If all committee members take their responsibilities seriously, the membership probably will find the recommendations worthy of support. And the organization will move effectively and efficiently toward its goals.

Self-conduct. Organizations often are judged on the behavior and appearance of their members. A group may be judged unfairly because of the behavior of a few members or even one member. A few youths who vandalize property can give a school a bad name. A first step in protecting an organization's good name is being especially careful when representing it or wearing its insignia. You may think you are acting for yourself, but the organization will be judged also. This includes following the considerations of manners and grooming that were discussed in Chapter 3.

The importance of proper conduct is seen in the fact that political parties often lose elections after a scandal that affects only a few members. It does not matter that most party members and nearly all of the leaders are honest and hardworking. The bad conduct of the few attracts the voters' attention and anger. Improper conduct by a few can wipe out the good efforts of many.

Responsibilities of Leadership

Fully carrying out your responsibilities as a member will prepare you for leadership. As a leader, you are still a member and still have all of those responsibilities. But now you also have added responsibilities. The responsibilities of informal leaders usually are not spelled out. Informal leaders are not officers of an organization, but are members who are respected by other members and who are listened to. Informal leaders can make sure that all points of view are presented by asking questions, expressing opinions, encouraging other members to speak, summarizing positions, and performing all the other roles that were discussed in Chapter 2. Their major responsibility is to work for the good of the group.

Formal leaders are charged with planning, with conducting meetings, and with coordinating the group's activities. The importance of their role makes it mandatory that care be taken in selecting them.

Selecting Officers

All members of an organization are responsible for electing capable officers. The key is matching the interests and abilities of possible leaders to the needs of the organization. It would not do to elect a treasurer who was poor at mathematics or a president who did not like to speak before groups of people. Matching the person to the job means that elections should not be popularity contests or snap decisions. They require thought and organization.

Officers should be elected who can work well together and with the membership. They need skill in group dynamics and human relations. Communication skills are particularly important. Officers should represent the values and beliefs of the members. A very important consideration is a willingness to serve; don't push offices on people who are reluctant.

Nominating Committee. Selection of officers requires a well-planned procedure. Nominating committees can ensure an orderly process. Such committees review the qualifications of people who are interested in serving and suggest others who might be qualified. Sometimes a qualified member might be willing to serve but reluctant to say so. A nominating committee can uncover such talent by asking qualified people if they would like to be officers. After asking potential officers if they want to serve, the committee presents a slate of officers to the membership.

The key to selecting good officers is matching their interests and abilities with the needs of the organization.

In choosing nominating committee members, the president should not include anyone who wants to be an officer. Committee members should be responsible members of the group and represent the different points of view of the members.

Nominations should not stop with the committee's recommendations. Further nominations should be called for from the floor. This helps open the selection process to all members of the group and may present qualified candidates who were missed by the committee.

The final step is election of officers. If there is a contest, a ballot will help ensure that votes of members are not unduly affected by others who are present. Members might be reluctant to vote for their first choice if they thought their friends disagreed. A show of hands or a roll call makes the vote of each member public knowledge.

Importance of Responsibilities

The officers of an organization hold a position of trust. By electing officers, members are saying: "We trust you to be responsible for our organization. We assume you know or will learn your duties and responsibilities." In a sense, officers are the glue that holds an organization together. For example, if the secretary does not keep an accurate record, decisions made by members cannot be carried out.

> James was treasurer of the Big Hollow FFA chapter, a very active organization. During the first half of the year, James kept careful and accurate financial records. He knew where every penny came from and where it went. Then, James began to grow bored with bookkeeping and the meticulous math needed for his assignment. He became lazy and sloppy. A number of the chapter's checks bounced. His carelessness had permitted the chapter to take on too many projects and to overdraw its checking account. A number of projects had to be stopped, and the chapter's reputation in the community was harmed.

James had not taken his responsibilities seriously. He had not been honest with himself and other members when he agreed to be treasurer. He wanted the recognition, but did not assume the responsibility and hard work that went along with accepting status.

General Responsibilities

If the leader does a good job in running meetings, the organization will be able to make decisions and to complete its business in an orderly way. Adequate planning and coordination make sure that all jobs get done on time.

If an FFA chapter plans a harvest dance, committee members should be assigned for the following functions: arranging a place to hold the dance, hiring a band, preparing refreshments, arranging for adequate supervision, and cleaning up after the dance.

The elected officers are responsible for making sure that each function is carried out, but not necessarily for doing the actual work, though there is nothing wrong with officers pitching in where needed. If even one task is neglected or done poorly, the success of the event can be endangered.

Leaders influence the quality of group actions. This means the leader is aware of the group's goals and tries to reach them through positive means. The leader encourages the group to act responsibly. In planning for a dance, some group members may not want supervision. Others argue that cleanup should be done by someone else. A responsible leader would try to persuade the group to accept such responsibilities.

Encouraging Participation. Group decisions and activities are most likely to succeed when all members are involved. The leader should make an effort to get everyone to take part in group activities. There is strength in numbers. Tasks can be divided into smaller and more manageable projects. A few people will not have to do everything. Leaders encourage participation by working to establish a pleasant and orderly atmosphere. People come to meetings that are well planned and productive; they will stay away if meetings are boring and disorganized.

Leaders also are expected to prepare and give reports. This task requires self-discipline to do the necessary work. You will need to develop your communication skills to write and deliver reports to members.

One of the most important duties of a leader is to ensure that group decisions are carried out. This requires checking with those responsible for doing the work and helping them solve problems. Maybe the FFA president knows more about suppliers of produce for an auction than a new member who has volunteered to check out suppliers. If the leader does not make sure the group's decisions are put into effect, activities may come to a stop.

Recognition for work well done will encourage members to do more. Praise and credit should be shared with those who deserve it. Leaders cannot do everything. Appropriate recognition ensures that there will always be enough hands to do the work.

Setting an Example

Leaders of organizations have a moral obligation to be positive and to practice responsible leadership. Members look to the leader to make sure that the organization's goals and purposes are carried out.

Responsible leaders identify and move to meet the accepted goals of an organization. Maybe the goals of an FFA chapter in a certain year are to provide community service through environmental projects. More general goals can be found in an organization's constitution, bylaws, and statements of goals and purposes. But not all goals will be written down. You will need to find out what the membership

Leaders have to encourage the participation of members in making sure that decisions are carried out.

wants. This can be done by asking individual members or by holding a meeting to discuss purposes and goals.

This does not mean you have to be a rubber stamp for the membership. You have a responsibility to remind members when they want to move away from the basic goals of an organization. Or the leader may need to work to help modify the organization's goals because of changes in the community or society. Leadership was required to open FFA membership to females, for example.

In seeking to change a group's goals or purposes, the leader should take a positive approach. What is to be gained from the change should be stressed rather than what is wrong with the old way. Information and persuasion are more effective than threats, ridicule, or personal attacks.

A leader may not agree with everything a group stands for, but should be comfortable with most of a group's goals. The leader is responsible for accepting and carrying out the desires of the membership. Accepting leadership means accepting the obligations to abide by members' wishes and to carry them out. The time to say an organization is moving too far away from its purposes is before a decision has been made.

Leaders are obligated to represent members' views accurately when speaking for the organization. A similar duty is not to use a group's name for purposes unrelated to its goals. It would be wrong to write a letter expressing personal views to a newspaper and use the group's name for identification.

Leaders also accept an obligation to prepare and improve themselves to do a better job. In the FFA this means attending and participating in leadership training conferences and other leadership training activities.

What a Leader Does

The effective leader knows how to plan, initiate, delegate, coordinate, make decisions, evaluate, and communicate. These acts are not necessarily performed separately. You may perform two or more at the same time.

Planning. Responsible leaders must first plan to use their own time efficiently. More than any other members of a group, leaders have many demands on their time. If leaders can plan their own time and activities, then they will have time to plan the group's activities as well. Planning helps a group function smoothly.

It also keeps the group from taking on more activities than it can handle. And good planning enables the group to use its

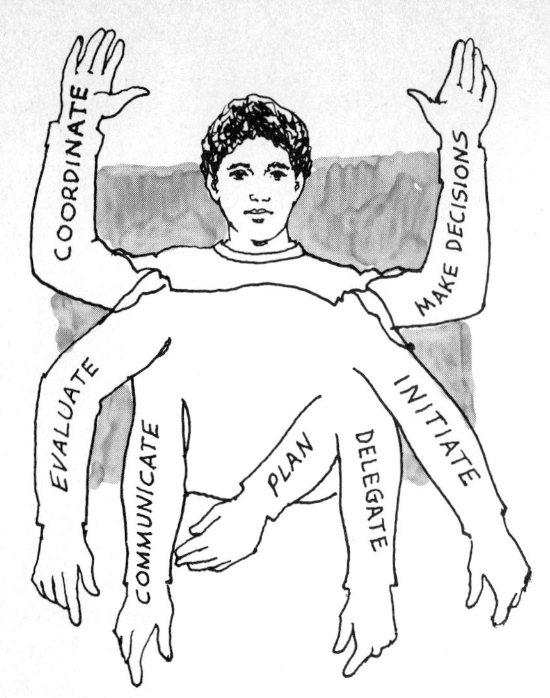

The effective leader has to use many talents in order to get the job done.

time to do the greatest possible number of activities. As the leader in the classroom, the teacher must plan carefully to ensure that all material in the course is taught during the time allotted. The teacher must be careful that a large part of the material is not left for the last few weeks of the year.

Initiating. This is the task that many people think of as leadership, but it is only one important aspect. *Initiating* means suggesting ideas to committees or to the membership. The leader can see the group's program and purposes as a whole. For example, an FFA chapter crop project might be a way to give members experience in soils, cooperation, and supervised occupational experience. The leader might see all of these possibilities better than members who are working on just one aspect of the chapter's program.

Delegating. No leader can do everything that is required of group leadership. Leaders must seek help from others. They must delegate both the responsibility and the authority for a job. If a group plans a picnic, different members would be asked to do different tasks. One person would be given the authority to arrange for the place to have the picnic. Another would be put in charge of food. A third would be asked to arrange transportation. The leader remains responsible for seeing that each person does the assigned task within the limits of the authority that was delegated.

Coordinating. This is one of the most important functions of leadership. Coordination ensures that the time and energies of the people in a group are used effectively. As coordinator, the leader picks the best person for each job. It is the coordinator's responsibility to make sure that the right number of people are used for each job and that different groups are not trying to do the same thing. Coordination involves deciding who is responsible for doing what and in what order.

If you were in charge of a farm construction business, you would delegate authority to the supervisors of different crews. You would make sure that each crew was busy every day. You would see that the foundation crew was sent to a job site after the excavation crew had done its work and before the carpenters who were to build the walls.

Decision Making. The leader must make decisions and help the group make decisions. Sometimes, leaders have to make unpopular decisions. A leader might have to decide to cancel an event, for example, if it was found the group did not have enough funds to carry it out. Leaders must keep the needs of the group in mind and try not to let the situation develop where too many unpopular decisions must be made. If the leaders keep attention focused on the budget, for example, events will not be planned that have to be canceled later.

To help a group make decisions, the leader must understand the situation in which the decision is to be made and the needs of the group and its members. What are the group's goals? What is it trying to achieve? What is the leader's sense of the proper way to proceed? Efforts should be made to enhance the group in the eyes of its members. Unpopular activities or decisions forced by officers can cause members to lose interest and to stop supporting the organization.

Evaluation. Leaders must constantly review an organization's activities to see if progress is being made in the intended direction. Do the activities advance the group's purposes? If the group planned a fund-raising event, did it result in more money? If not, what else needs to be done? Is something being done wrong? Leaders use formal and informal evaluations. Informal evaluation is done from day to day. It may consist of the leader's sense of how things are going based on what people say and what the leader can see. A more formal evaluation would be a periodic check of the program of activities to see if initial plans for the year were being completed.

Communication. Without communication, none of the other functions of a leader can be completed. It will do little good to plan if the plans cannot be communicated to others. Coordination is impossible without communication. Communication should get special attention from leaders. Each act of leadership requires that messages are sent and received accurately, as was discussed in Chapter 7.

Member and Leader Responsibility: A Review

Learning to assume responsibility as a member of a club or an organization prepares you to work more effectively as an employee, a citizen, and a community member. The more you develop your abilities as a leader, the better prepared you will become to lead your own life. Leaders are members who have taken on a greater amount of responsibility. The skills needed by leaders are the same ones needed by an employee, a family member, or a citizen. Everyone needs to be able to work with others, to plan, to communicate effectively, to make decisions, and to evaluate goals and activities. Every organization, to run effectively, needs members and leaders who take their responsibility seriously.

THINKING IT THROUGH

1. Why is participation necessary for the functioning of an organization?
2. How does participation in organizations influence the effective functioning of a democratic society?
3. What are your responsibilities as a member of an organization?
4. Why may member responsibilities include conduct and appearance of members?
5. How can capable officers be elected?
6. What are some characteristics of responsible leadership?

CHAPTER 12

THE FFA: A LEADERSHIP LABORATORY

Jason and Jonah were good friends and excellent students. Both excelled in classwork, took an active part in classroom discussions, and did well on tests. Jonah also became active in several school organizations. He became president of the FFA chapter in his senior year.

Jonah often urged Jason to join some of the groups. He tried to explain how important it is to learn to be an effective and responsible member of an organization. But Jason refused. He argued that joining organizations would take away from his main purpose at school. "I'm here to get an education so I can get a good job," he insisted time and again.

After graduation, both joined the same agricultural supply firm. The manager liked to hold employee meetings to talk about ways to improve the business. Jonah contributed many ideas at these sessions and soon was receiving promotions and pay raises. Jason had good ideas too, but he felt uncomfortable expressing them before a group of people. His ideas often were offered later by other employees who got the credit and the rewards.

Jason learned too late that education is more than book learning and classroom discussions. He had had dreams of becoming a leader while in school and had studied about leadership. But he had never practiced doing the things he had learned.

Studying about leadership and personal growth is not enough. You need an opportunity to apply the knowledge before you can make it fully your own. Learning about leadership without trying it is like reading about music without practicing on an instrument. Like music, leadership and effective participation in groups require a lot of practice.

Learning is an active process. Ralph Ellison, a novelist, wrote: "Without the possibility of action, all knowledge comes to one labeled 'file and forget.' " Vocational education stresses the use of a laboratory to develop and perfect skills and abilities that were studied in the classroom. Mere theoretical knowledge is not enough. Just as you need to work with live animals and real soil to gain knowledge and skills in livestock judging or soil science, you need to try your leadership knowledge and skills with other people.

This practical side of learning is called laboratory learning. To learn about small engines, you actually take one apart and put it back together again. To learn parliamentary procedure, you seek the floor, make motions, or preside during the meetings of a group.

As a student of leadership and group process, you need a laboratory in which you can perform and practice the knowledge, roles, and skills you study in the classroom. The FFA is the laboratory you need to gain practical experience in membership, leadership, group dynamics, and citizenship. What you learn in this laboratory now will be invaluable when you seek a job later on.

CHAPTER GOALS

In this chapter your goals are:

- To analyze how the FFA serves as a laboratory for leadership development
- To identify the types of recognition available to FFA members
- To explain how the FFA program has developed
- To identify major FFA activities
- To evaluate the personal advantages of becoming a member of the FFA

The Leadership Laboratory

A student organization fits the description of a laboratory for participation and leadership. This is so because much of your life will be spent with one organization or another. Your workplace will be an important organization where you spend time making a living and finding personal fulfillment. An *organization* is a formal group as defined in Chapter 6.

At this point in your life, you may need to join more than one organization to take care of different interests. You might seek one that is related to your vocational and career interest. This would be an intracurricular organization that ties classroom learning to the laboratory-organization setting. *Intracurricular* implies that the organization is a *part* of the school and its programs, not an *addition* to those programs. Vocational youth groups are intracurricular. The FFA was the first such organization and has served as a model for others.

The advisor to vocational youth organizations is also the classroom teacher. The vocational agriculture teacher serves as advisor to the FFA chapter. The FFA was developed to provide practical experience for what is learned in vocational agriculture courses, as well as opportunities and challenges for personal growth and leadership development.

The FFA provides many activities and contests that will help you practice the skills required for effective leadership. Activities include parliamentary procedure, public speaking, community services, and citizenship. The FFA also has a comprehensive leadership training program. To strengthen the vocational aspect of your course work, the FFA degrees and proficiency awards recognize progress in leadership, personal development, and supervised occupational experience programs.

The FFA was developed to serve the specific leadership and personal-development needs of students in vocational agriculture. It is a natural and unique laboratory for you. The FFA was established specifically for all students of agriculture. They are now found on the farm, in agricultural supply firms, and in many related occupations. There is a wide variety of opportunities for participation in the FFA for students who live on farms, in towns, in suburbs, and in cities.

FFA Program

The FFA's program emphasizes the recognition of achievement by members. The FFA believes that members will be moti-

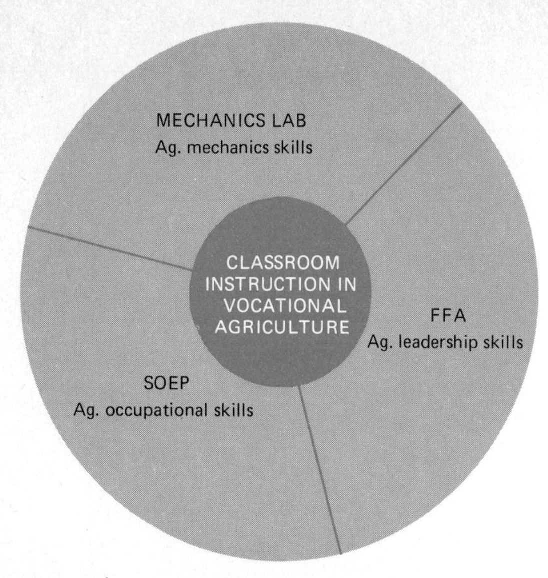

Laboratories in vocational agriculture.

vated to participate actively by a system that rewards them from the local to the national level. Members are thus encouraged to try to achieve in areas related to participation, leadership, and work experience. Recognition includes contests with other members and awards determined by fixed standards.

Degrees and Awards

Degree Program. The FFA degree program is based on the members' supervised occupational experience (SOE) program and leadership development. Degrees are directly related to accomplishment. There are four degrees for FFA members: Greenhand, Chapter Farmer, State Farmer, and American Farmer. Each is awarded by the appropriate level of the FFA organization.

The local chapter awards the Greenhand and Chapter Farmer degrees. To qualify for the first degree, which is the Greenhand, a student must be enrolled in a vocational agriculture course and have a satisfactory plan for supervised occupational experience. The student also must learn and explain the FFA creed, motto, and salute; describe the emblem, colors and symbols; explain the use of the jacket; have a knowledge of FFA history; know the duties and responsibilities of an FFA member; have access to an FFA *Official Manual*; and apply for the degree.

To earn a Chapter Farmer degree, the member must have received the Greenhand degree and meet additional requirements as outlined in the national FFA constitution (found in the FFA *Official Manual*) and in the local constitution. The focus is on progress in leadership and personal development, supervised occupational experience, and FFA activities.

The minimum requirements for the State Farmer degree and the American Farmer degree are also outlined in the national FFA constitution. They recognize continued progress and success in FFA membership, leadership development, personal growth, vocational agriculture course work, supervised occupational experience, and community activities. Not everyone qualifies for State Farmer and American Farmer degrees. Each state can award State Farmer degrees to up to 3 percent of that state's membership. American Farmer degrees also are awarded on the basis of the number of state members.

Proficiency Awards. Outstanding students may qualify for proficiency awards when they have excelled in progress toward a career and in leadership development. Proficiency must be demonstrated in one or more agricultural occupations. The member must show actual mastery of the techniques and skills of dairy production, for example. Over 20 proficiency awards are available in such areas as agricultural sales and service, outdoor recreation, and crop production. Awards may be made on a chapter, state, and national basis in each of the areas.

The FFA gives out many awards and degrees such as the plaque here being given to a local farmer.

Achievement Awards. This awards program recognizes FFA members for accomplishments in the instructional program and in FFA leadership activities. Members compete against a standard set of performance criteria rather than against each other. The program is designed to involve more students in the FFA awards program and to give every student more chances to achieve recognition. Awards are made in such areas as vocational skills, leadership activities, career knowledge, and safety practices. The local chapter grants the awards.

Experience and Rewards

Your FFA chapter provides you with the opportunity to learn, to grow, and to serve. It also gives you a chance for financial gain by encouraging earnings through SOE programs. In addition, FFA members may compete for recognition and awards on a local, area, state, and national basis. These contests include competition in areas such as public speaking, extemporaneous speaking, and parliamentary procedure. The contests help members in their personal growth and leadership development.

Other contests are related to agricultural occupations. These include agricultural demonstrations. A member could, for example, set up a demonstration in agricultural sales. Actual merchandise would be on hand to be sold. A panel of judges would evaluate each member's ability to sell, including personal appearance, sales techniques, and knowledge of the items being sold.

Contest judging is related to classroom instruction in such areas as livestock, land, horticulture, poultry, and meats, and tests members' technical knowledge. In addition to providing recognition for those who take part, the contests give students an opportunity to develop knowledge and skills. Details about these and other contests may be found in the FFA *Official Manual* and *Student Handbook*.

The FFA program also provides members with a chance to travel, which can be a rich learning experience. Field trips, conferences, and conventions give members a chance to meet other people and to learn firsthand the way things are done in other areas. Members also have the opportunity to learn to demonstrate their knowledge and skills in different settings.

With members planning and carrying out the activities of local chapters, the FFA serves as preparation for life. It is as much a laboratory for vocational agriculture as the shop, the greenhouse, or the SOE program. Active participation in the FFA helps vocational agriculture students make the most of their learning experiences.

The FFA provides many of the same opportunities as athletics for developing personal abilities. Indeed, the FFA has some advantages over sports. For example, it provides recognition for many rather than for a few, and it emphasizes development of a greater range of skills.

AWARDS

All Agriculture Proficiency Awards are presented to Individuals

AWARD AREA	LOCAL	STATE*	REGIONAL**	NATIONAL***	SPONSOR
Agricultural Electrification	Metal	$100 Check and a Certificate	$250 Check and a Plaque	$500 Check and a Plaque	Edison Electric Institute
Agricultural Mechanics	"	"	"	"	International Harvester Company
Agricultural Processing	"	"	"	"	Carnation Company
Agricultural Sales and/or Service	"	"	"	"	Allis-Chalmers Corporation
Beef Production	"	"	"	"	Nasco and Sperry-New Holland
Crop Production	"	"	"	"	Producers of Funk's G-Hybrids
Dairy Production	"	"	"	"	Celanese Chemical Company
Fish and Wildlife Management	"	"	"	"	American Fishing Tackle Manufacturers Association
Floriculture	"	"	"	"	General Fund
Food and/or Vegetable Production	"	"	$200	$250	General Fund
Forest Management	"	"	$250 Check and a Plaque	$500 Check and a Plaque	Weyerhaeuser Company Foundation
Home and Farmstead Improvement	"	"	"	"	Upjohn, TUCO, Asgrow and Cobb Organizations, Agricultural Division of The Upjohn Company
Horse Proficiency	"	"	"	"	The American Morgan House Foundation
Livestock Production	"	"	$200 Check and a Plaque	$250 Check and a Plaque	General Fund
Nursery Operation	"	"	$250	$500	Weyerhaeuser Company Foundation
Outdoor Recreation	"	"	$200 Check and a Plaque	$250 Check and a Plaque	General Fund
Placement in Agricultural Production	"	"	$250 Check and a Plaque	$500 Check and a Plaque	Hesston Corporation, Inc. and Shell Chemical Company
Poultry Production	"	"	$200 Check and a Plaque	$250 Check and a Plaque	General Fund
Sheep Production	"	"	"	"	General Fund
Soil and Water Management	"	"	$250 Check and a Plaque	$500 Check and a Plaque	Ford Motor Company Fund
Swine Production	"	"	"	"	Pfizer, Inc., Agricultural Division
Turf and Landscape Management	"	"	$250	$500	O.M. Scott & Son Company

*The number of state awards available from the National FFA Foundation varies by state.

**All Regional Winners must attend the National FFA Convention to receive their award.

***All National Winners must attend the National FFA Convention to receive their award.

The FFA proficiency awards.

An FFA livestock judging contest.

Planning FFA Activities

"The primary aim of the Future Farmers of America is the development of agricultural leadership, cooperation, and citizenship," states the FFA *Official Manual*. An important part of the learning laboratory experience of the FFA is the cooperation and citizenship skills required of members when they plan their own chapter program of activities. This program is established each year and is built around the needs of the individual and the community. The program indicates the direction to be followed to reach a definite set of goals.

All members serve on committees and take part in the discussions that plan the program of activities. The program evolved by the membership becomes the plan and guide for the year's activities. The committees that develop the chapter program of activities also make sure that all activities are completed. They are organized to include the following eleven areas.

Supervised Occupational Experience. This activity encourages students to apply practically what they learned in the classroom. Activities depend on where the chapter is located. Supervision in dairy production might not be appropriate in a city FFA chapter. But there could be urban programs that emphasize agricultural processing, sales, landscaping, or nursery operation. Specific activities could include establishing a chapter loan program, sponsoring a pesticide safety program, or promoting work experience in agricultural professions.

Cooperation. Cooperation among members helps determine the success of a chapter. These activities are designed to teach members to work together on projects, either with other members or with outside groups. Activities could include establishing a chapter farm, cosponsoring a Farm-City Week with a local civic club, or sponsoring a children's zoo at the local fair.

Community Service. Community Service activities help members learn and practice good citizenship and leadership abilities needed now and in the future. The focus is on making the community a better place in which to live and work. A chapter may complete an activity by itself or join with other groups on a larger project. A chapter might study the need for a bike trail and work out the suggested route. Another example is providing for the maintenance and repair of equipment in a city or county park.

Leadership Development. Becoming a leader takes time, effort, study, practice, and a chance to learn from mistakes. FFA chapters provide members with opportunities for all of these. Daily chapter operation provides many chances for leadership. Members can try out the various roles that were discussed under "Roles of Group Members" in Chapter 6.

The chapter also can provide specific

activities that enhance development of leadership abilities. These include appointment of all members to committees; mock meetings at which each member gets a chance to practice different leadership skills; demonstrations of parliamentary procedure for classes or other groups; and providing members the chance to take part in FFA-sponsored public speaking contests.

Earnings, Savings, Investments. Every FFA chapter needs funds to pay for its activities. Fund raising gives members a chance to use their membership and leadership abilities and to learn how an organization manages money. Fund-raising events may be coordinated with local merchants who want to help service-oriented school organizations. Activities could include selling pine seedlings for reforestation, repairing and selling used farm equipment, and giving special recognition to thrifty members.

Conduct of Meetings. This activity offers many opportunities for developing leadership skills. It also is essential to the efficient conduct of the chapter's business. Everyone can learn to practice correct parliamentary procedure, such as the proper form of obtaining the floor, making a motion, or presiding over a meeting.

Members also can practice public speaking and communication with groups. Different members could be asked at each meeting to give a report on some aspect of agriculture. Outside speakers can be invited. Members can take turns introducing the speakers. The chapter could make a practice of asking a member to speak extemporaneously at each meeting.

Scholarship. Scholarship includes activities as well as classwork and good grades. The FFA chapter can encourage good scholarship by widening the horizons of members. An agricultural marketing project provides concrete experience that makes classwork more real and meaningful. You may associate with students who have other areas of interest and find new challenges for yourself. Other projects might include posting a chapter honor roll and recognizing members with good grades in agriculture.

Recreation. Recreation gives members a chance to practice some of the social skills that were discussed in Chapter 3. Recreation also can extend and broaden the learning experiences of students. Chapters can organize sporting events, dances, dinners, picnics, hikes, competitions, and field trips to points of interest. Often an agricultural theme can be stressed.

Public Relations. Public relations concerns both actively achieving your goals and telling others about your accomplishments. It also means that members of the public can be involved in FFA activities. All these aspects would be combined in inviting the mayor to speak to the chapter on the relationship of agriculture to the life of the community. Members would prepare and distribute news releases to newspapers, radio stations, and TV stations. One release would announce the event and a second would contain an account of the speech.

A weekly news column on FFA activities can be prepared and distributed to the media. When members obtain special recognition, news releases can be accompanied by a picture. Recognition luncheons or breakfasts for faculty members or local agricultural merchants can be organized during FFA week.

State and National Activities. FFA members can take on greater challenges and responsibilities by taking part in state

FFA chapters offer a varied program of activities.

and national activities. These include degrees, proficiency awards, contests, international programs, and officeholding. Activities include selection of local chapter delegates and representatives to attend state and national conventions.

Alumni Relations. Many former FFA members want to stay involved in some way. This permits them to share their experience with younger FFA members and to serve their school and community. Alumni relations benefit student members by keeping them in contact with experienced members of agricultural industry. This may open a door to employment in the future. The alumni can be called on to help the chapter within their specific area of experience. They can be asked to serve as judges of FFA contests. Alumnus-of-the-Year awards can provide recognition for outstanding former chapter members.

Citizenship Training

FFA activities strongly emphasize learning and personal development for responsible citizenship. The Building Our American Communities (BOAC) award program is designed to develop interested, experienced, and knowledgeable community leaders and citizens. All members try to create better communities that provide services, job opportunities, and a better quality of living. The BOAC program seeks to develop a sense of pride in the community and initiative in making it a better place to work and live.

The BOAC program offers FFA members a chance to study and work with the local political structure and other community organizations. Students study the community's economic, social, and environmental needs and learn how community support is developed for key programs. Members actually work with community leaders to develop and carry out long-range programs in such areas as safety, soil conservation, pest control, or community beautification.

Group Leadership Activities. The FFA sponsors special group leadership programs. It includes camps and conferences where leadership skills can be developed fully. Also included are leadership schools; training seminars; leadership tours; area, state, and national conventions; the Washington Conference program sponsored by the National FFA Organization; and the National FFA State President's Conference. While all these activities promote good citizenship, the Washington Conference emphasizes this aspect. Members can observe the federal government in action and meet many of the people who make it work.

All of the activities provide for training and an exchange of ideas among members. Participation by a few members in one of these events can benefit an entire chapter. The activities are designed to develop leadership for chapters, to develop potential community leaders, and to create an atmosphere for the exchange of ideas among chapters and FFA leaders.

Your own chapter can arrange leadership tours. You could visit and observe your local government in action. A visit could be arranged also to state agencies and the state legislature. The purpose would be to study how leaders carry out their daily responsibilities.

Citizenship training is another important service that the FFA provides to its members. Through these FFA programs, you begin to learn more about the community where you may work and raise a family. You begin to know people who will have an important influence on your life. The training will be useful in carrying out your citizenship responsibilities, in your occupation, and in your work with community civic organizations.

FFA Background

The FFA was founded to serve as a leadership laboratory for agricultural students. It is an integral part of the agriculture curriculum. What is learned in the classroom can be applied in chapter activities. Work in one reinforces work in the other. Teachers of agriculture serve as advisors to the FFA chapter. The FFA also provides for a broad scope of state and national activities.

FFA History

Vocational agriculture courses were established in 1917 under the provisions of the federal Smith-Hughes Act, which provided for vocational education. Local agricultural clubs developed in a few years. They were mainly social and recreational, but included some educational features. These groups formed because of common backgrounds and needs. The Future Farmers of Virginia provided leadership in the development of local and state organizations.

Many leaders began to dream of a national organization of agricultural clubs. This dream became real in 1928 when the FFA was organized in Kansas City, Missouri, with eighteen states represented. Leslie Applegate of New Jersey was the first national FFA president. Dr. C. H. Lane of Washington, D.C., who was responsible for supervising agricultural education programs at the national level, became the national adviser.

Thirty-three states sent representatives to the second convention. The FFA creed was adopted at the third convention, along with the FFA colors, national blue and corn gold. In 1939 the national FFA bought 28.5 acres [11.5 hectares] of land that had been part of George Washington's estate to establish the National FFA Camp. The National FFA Foundation was founded in 1944 to help finance the awards program. Over $750,000 was contributed to the foundation in 1976.

The National FFA Supply Service started in 1947, and National FFA Week was begun the next year. A federal charter was granted in 1951 under Public Law 81-740. In 1952 the Official Code of Ethics was adopted and *The National Future Farmer* magazine was started. In 1959 the first National Leadership-Citizenship Conference was held in the nation's capital.

In 1969 delegates to the national convention named the first Star Agribusinessman. The Building Our American Communities program was started in 1970. When Alaska received a state charter in 1976, it meant that each state and Puerto Rico had a state association.

The FFA emblem.

Not all of FFA's half million members live on farms or in farm communities. Modern agriculture includes a wide range of occupations, and these are reflected in the FFA's activities. You have the opportunity to develop your leadership abilities at the local, state, and national levels. Students serve as officers at all three levels. There also is an adult board of directors for the national organization.

The FFA and You: A Review

The strong stress the FFA places on the preparation for jobs and personal growth makes it an ideal program for vocational agricultural students. Its *Official Manual* lists twelve major purposes. One of them is to strengthen the confidence of students of vocational agriculture in themselves and in their work. By tying its program directly to the vocational agriculture curriculum, the FFA offers direct experience that will be useful when you seek full-time work. In addition, the practice you get as a member and leader of the organization will help make you a productive employee in any area you choose to work. The broad program of occupational, personal, leadership, and citizenship development prepares you for the rest of your life.

THINKING IT THROUGH

1. How does the FFA serve as a laboratory for leadership experiences?
2. What types of recognition do FFA members receive?
3. What are the requirements for the Greenhand degree?
4. What is the FFA proficiency awards program?
5. What leadership development activities are available for FFA members to participate in?
6. How are FFA activities planned?
7. Why and how has the FFA developed since 1928?

CHAPTER 13

LEADERSHIP, PERSONAL GROWTH, AND WORK

Stan enrolled in vocational agriculture to prepare for a satisfying career. He knew he wanted a job after high school. Completing his supervised occupational experience in a nursery during his senior year convinced Stan he would like horticulture. He found an opening in a garden supply center through a newspaper ad.

First, Stan visited the garden center. He noticed that it was pleasant and cheerful. The sales staff was courteous and friendly. They gave advice about plant care to many regular customers who came into the store. This was a place where something could be learned.

Stan sought out the manager and asked for an interview: ''I saw your ad for a salesperson in the morning paper. When might I be able to arrange for an interview?'' He was told to come back at 9 a.m. the next day. He used the evening to prepare a personal data sheet and to think about what might be asked. Next morning, he dressed carefully and arrived a few minutes early for the interview.

During the interview, Stan sat up straight, looked directly at the interviewer, and answered all questions clearly. ''Why do you want to work here?'' Stan was glad he had visited the store. ''The atmosphere is friendly and I think I could learn a lot. I enjoy working with plants.'' He was called a few days later and told the job was his.

Getting a job requires convincing the employer you can do it. Employers want to know that you are motivated, enthusiastic, and willing to work and learn. School performance, activities, and part-time work are important. They show the employer what an applicant can do.

The purpose of personal development is to improve the ability to be a responsible member of an organization, especially the one that employs you. Current efforts to grow and change prepare you to do good work for an employer and for community service.

CHAPTER GOALS

In this chapter your goals are:

- To identify and evaluate job openings in light of your personal desires and abilities

- To follow correct procedures in applying for and interviewing for a job
- To establish and maintain successful relations with supervisors and coworkers
- To analyze the characteristics that lead to advancement on the job

Finding a Job

Many people don't know how to find or get a job. The ability to do both is important because so many people change jobs several times during their lives. Getting a job becomes manageable if it is taken a step at a time. Some work must be done to get work.

Deciding on a Job

The three basic steps in deciding on the kind of work you want are analyzing yourself, identifying the possible jobs, and making the decision.

Personal Factors. Self-knowledge is a must when thinking about a job or career. People often make mistakes because they don't think through the personal factors. They follow parents' wishes, go along with friends, or take the first job they find. It never occurs to them to think through what they want for themselves.

It's hard to find a job that fits you exactly, but you can try for one in which you can grow, learn, and enjoy yourself. Before looking for a job, weigh carefully your background, preparation, desire for more education, preferences, and needs. Write everything down.

Background includes family situation, work experience, activities such as FFA, hobbies, and likes and dislikes. When developing this list, think how each item could affect your choice of work. How are your interests reflected in your activities

To decide the kind of work you want, analyze yourself, make personal contacts to identify potential employers, and make a decision by learning about a job from an employer.

and hobbies? If you like to hike and ski, you probably would like outdoor work. What is your financial situation? Must you get a job right away?

Additional Preparation. Learn what preparation is needed for your job or career choice. Then find out where it is offered. Maybe you can get it on the job or at night school. Check out the many different kinds of institutions that offer education after high school. If you can't afford to go to one now, maybe you will later.

Write down all the preparation you have already. Part-time work on a dairy farm is preparation in agriculture. FFA activities provide education in leadership, public speaking, and many other areas. High school courses are another form of education. An agricultural mechanics course offers preparation in farm equipment repair or carpentry.

Likes and Dislikes. Don't skip this area. You will spend a lot of time on the job. What challenged or bored you in school or previous work? If you disliked chemistry lab, agricultural research may not be for you.

> Alonzo liked the outdoors. He enjoyed mountain climbing and wanted to live in the West. He liked physical work and work with fewer people. He thought a career as a forester would fill most of these needs.

Thinking through and writing down facts about personal background, education, work experience, and likes and dislikes can be the beginning of a personal employment profile. What kinds of jobs fit this profile?

Identifying Jobs. Job information can be found in many places. Direct contact with a manager or employment office of a firm quickly determines if any jobs are available. A daily check of the help-wanted ads in the local paper shows more generally what jobs are available. Employers list jobs at the school placement office or in local offices of the state employment agency. And there may be other public or private employment services in your community. Private services may charge a placement fee. Personal contacts—teachers, parents, friends—also know about jobs.

Making the Decision. Drew Hammond completed all the steps in identifying what he wanted to do. Then he found an opening on the sales staff of a farm cooperative. He was tempted to rush and apply for the job. But he knew he might be sorry for a decision made too quickly. First, he wanted more information. He called to set up the interview. He asked a friend who worked at the co-op what the sales job involved. It meant work in the stockroom as well as selling.

Drew checked on the firm's reputation. Customers told him the co-op was honest and fair. His friend told him about wages and fringe benefits. They compared well with what other employers paid. There was an insurance plan, sick leave, paid holidays, and a paid vacation. However, advancement was slow. There was not much turnover among the employees.

Learning about the job and employer put Drew in a good position to decide about the job. He was ready to make a realistic decision based on facts. He was willing to accept slow advancement because other factors were positive. He also knew the decision could be changed later. He wouldn't be stuck with a job he didn't like forever.

Now Drew could follow the decision-making steps outlined in Chapter 4: define the problem (do I want the job);

identify the facts (pay, fringe benefits); plan for possible solutions (take this job/look for another); evaluate the alternatives (good solid job now/finding another takes time); and decide (I'll take this job).

Getting the Job

Once a job is located, you must convince the employer to hire you. Before the job interview, you must contact the employer and prepare a personal data sheet.

Personal Data Sheet. This is an introduction to the employer. Often they ask for personal data sheets, or resumes, first and then select the persons to interview on the basis of the most promising sheets. A good sheet won't guarantee a job, but a bad sheet can lose a job. Personal data sheets should be neat, accurate, and typed. Sloppiness makes a poor impression and could cost a job. The illustration on page 141 shows the type of information that should be included.

Information on the personal data sheet should be as specific as possible. Specific skills mean more than a job description. In listing activities, be sure to include all offices, responsibilities, or awards.

Contacting the Employer. Contact may be in person, by phone, or by letter. If the employer wants to be called, the telephone procedure described in Chapter 7 should be followed. Organize your thoughts and speak clearly.

> Hello, Mr. Erickson. My name is Drew Hammond. I want to apply for the job that was advertised in yesterday's *Chronicle*. I have some experience working with people. I would like an appointment for an interview.... Thank you, I'll be there at 9 a.m. sharp.... Good-bye.

If the employer gives you a job application, look it over first. Find information you don't remember before filling it out. This may include health records or dates of previous employment. Follow the instructions and fill in all the blanks neatly. If some items don't apply to you, write "not applicable" in the blanks.

The Job Interview

The interview is your chance to persuade an employer to hire you. Careful preparation is required to make the best possible impression. Try to think through questions that might be asked. You may be asked about yourself or the company. Be prepared for surprise or intimidating questions that are designed to test your reactions.

Here are some questions that are often asked: What can I do for you? Why do you want to work here? Tell me about yourself. What are your goals in life? What do you want to be doing 10 years from now? Are you ever late?

Try to understand and respond from the point of view of the interviewer, who is seeking the best person for the job. Don't lie or twist the truth to say what you think the interviewer wants to hear. But emphasize your strong points. And be prepared to admit that you don't know or have not decided something yet.

Consider the convenience of the interviewer. It is best to go alone. Never be late. And show you care about the interviewer and yourself by being clean, neat, and well-groomed. Sloppiness is a signal of how people value themselves. An interviewer will not value someone who is not clean and neat. Take along an extra copy of your personal data sheet and a pen or pencil.

Personal communications skills (Chapter 7) will come in handy during the

Drew Hammond
Box 129
Williamson, Wis.
941-3265

Age: 18
Health: Excellent
Born: 9/17/59

Education: Williamson High School, graduated 1976. Vocational agriculture courses. High B average. Cooperative education experience in corn production and farm equipment maintenance.

Activities: Williamson FFA Chapter. Vice President in senior year; State FFA Degree; Proficiency awards in crop production and agricultural mechanics. Secretary, Williamson High School Senior Class and Junior Class representative on the Student Council.

Work Experience: 1975-76: Co-op education and summer work, Golden Nugget Farm.
1974-75: Co-op education work, part-time farm equipment repair, Williamson Farm Supply Center.
Summer 1974: Selling and care of plants and garden supplies, Grassy Knoll Nursery.

References:

Anton Granik, Manager, Golden Nugget Farm, Box 22, Williamson, Wis. 941-3354.

Stewart Loomis, FFA Advisor, Williamson High School, 74 Water Street, Williamson, Wis. 941-8842.

Gertrude Braun, Head Teller, First National Bank, 28 Main Street, Williamson, Wis. 941-6660.

A sample personal data sheet.

Remember to prepare carefully for an interview.

interview. Remember, you communicate by gesture, body posture, tone of voice, and facial expression, as well as by words. If your statements and tone of voice are flat or unemotional, you won't sound very enthusiastic. You may be a little nervous. This is natural. Careful preparation will help hold nervousness to a minimum. And it's a poor policy to smoke or chew gum during an interview.

Introduce yourself, stating your name clearly. Shake hands if a hand is offered and sit when asked. Look directly at the interviewer when talking. Use correct grammar and avoid slang. Sit erect and avoid nervous gestures like rubbing your face, pulling your ear, or smoothing your hair. Seek to convey an alert self-confidence.

Answer all questions carefully, directly, and honestly. Don't overrate or underrate yourself. Watch out for questions like, "Do you ever miss work?" A quick "No" probably will be interpreted as a lie. A more appropriate answer could be, "Not unless I am ill or have arranged to be absent ahead of time."

Stress strong points, be willing to admit weak ones. But express your willingness to improve weak areas. Offer evidence of this willingness: "I know my public speaking ability needs work. That's why I took a speech class and entered FFA public speaking contests."

Ask questions about the job or the company. You can ask about fringe benefits, chances for advancement, and training opportunities, for example. Watch for a signal that the interview is about to end.

Plan what you will say to close the interview. You can end an interview with an expression of thanks and of strong interest in the job.

After the Interview. You may not be offered the job on the spot. If you are, you may ask for a few days to think over what you learned in the interview. Ask when a decision is needed and promise to call with your decision by that date. Follow the interview with a short thank-you letter that restates your interest in the job and offers more information if it is needed.

If you don't hear from the interviewer within a reasonable time, you may call and ask about the outcome of the interview. If the interviewer gave you a specific time period for the decision, wait until that expires before calling. Ask if you are still being considered and again express your interest in the job.

Going to Work

Whether you are self-employed on a farm or work in a business, you will work with people. The workplace may be the most complicated place in your life. You will spend a lot of time on the job. Depending on the type of job, you will have interactions with coworkers, supervisors, middle and top management, customers, and suppliers.

The need to get along at work is a major reason for concern with human relations, group dynamics, personal and group communications, leadership, and personal growth. Learning to work with groups and with other people is invaluable on the job. Workers must be aware of their own needs and that coworkers have similar needs—for self-esteem, personal fulfillment, friendship, and approval.

Most workplaces are organizations much like FFA. A business may be larger and more complicated, but it demands the same skills, attitudes, and behaviors. The first few days at work may be confusing. This is to be expected. Everyone has the same experience.

Keeping a Job

The first requirement of a job is the ability to get along with others. Getting along with coworkers is much like getting along with classmates. They expect you to be dependable, loyal, helpful, and understanding. Emphasizing these characteristics will bring friendliness in return. You will find an occasional person who refuses to be friendly. But remember, you can't be friends with everyone.

Employers want employees who are dependable, honest, and reliable. They will be satisfied if you get to work on time, don't extend breaks, and learn your job and do it without constant supervision. Customers expect courtesy, helpfulness, and honesty. The key to success on the job is getting along with others. Try treating others as you like to be treated.

Advancing on the Job. The first step to a promotion is to accept each assignment and do it well. Some tasks may seem boring. They become less tedious if you realize they won't last forever and that doing them well may help bring a promotion or pay raise.

While doing your basic job, be prepared to upgrade your skills. What skills are needed for a specific job or for advancement? A lot can be learned by watching, listening, and asking questions. Most people are happy to share what training is needed for a job.

> Kurt wanted to advance in the large tractor dealership where he worked as a mechanic. He knew man-

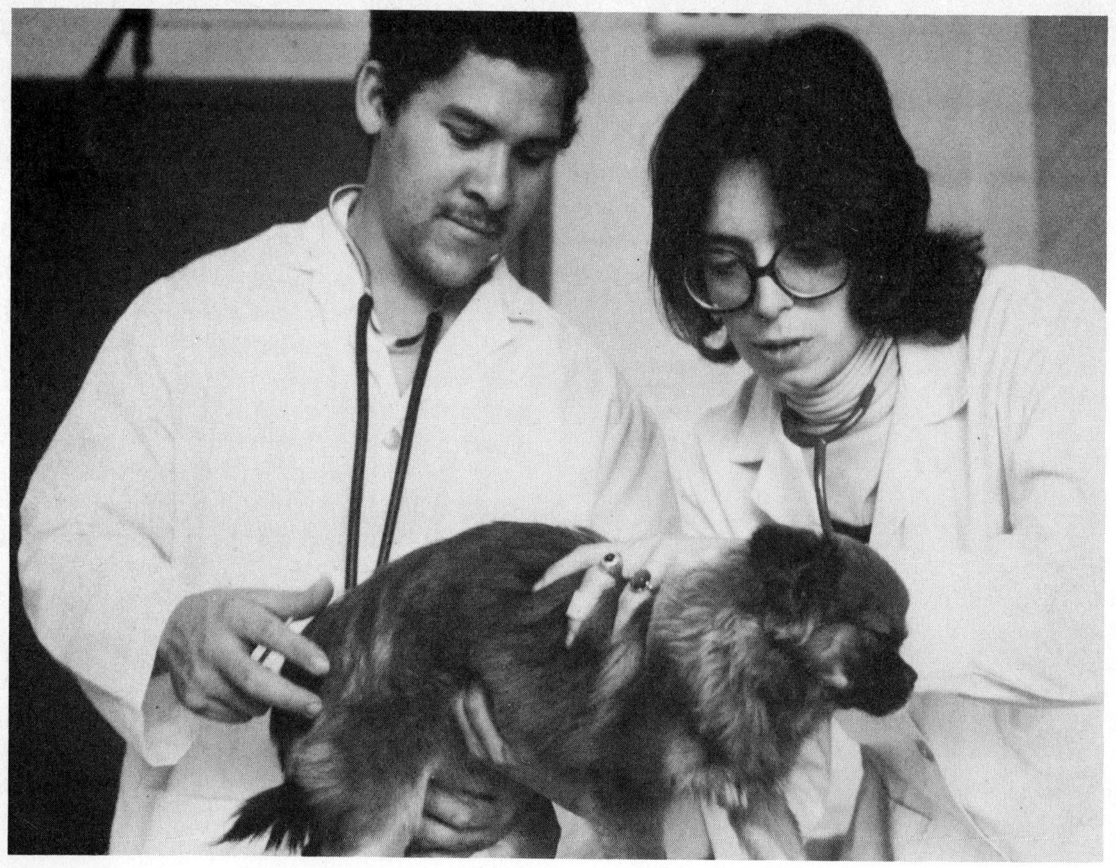

Working well with others is the first requirement of a job.

agement skills were important for higher positions. So he took some night courses in management. At the same time, he applied for the management-training program offered by the employer.

He also noticed that the top managers were able to speak effectively. Kurt took a speech class and made a point of discussing repairs with the customers. He also joined the Rotary Club and volunteered for committee work. When he was named chairman of a program committee, he got a chance to practice his skills at presiding over meetings. Mr. Ross, the firm's manager, was skilled at running meetings. They were a bit more informal than the FFA meetings Kurt remembered, but there were a lot of similarities between them.

Kurt watched for his opportunities. He applied for openings when they were posted on the bulletin board or mentioned in meetings. He missed out on promotions the first couple of times. But his supervisor praised and encouraged him. Eventually, Kurt got a promotion to supervisor of the stockroom.

The Day-to-Day Routine. There are good days, bad days, and in-between days at work. Skill in human relations is needed for every situation, both the day-to-day work and the hard times.

After a while, it is easy to get careless. What difference does it make if you get to work a few minutes late? Who cares about dress and grooming? Maybe work can be slacked a little.

These attitudes will be noticed. They may not cost a job right away. But in a crisis, the first to go will be those who are less interested. They also are the last ones to get promotions. It is easy to get careless and then complain when passed over for a promotion.

Learn your job well and look for places to show initiative. Maybe there is a way to store inventory that requires less space. It could be something simple like placing the water cooler in a more central position. Learn and practice company policies. They were developed because they are effective.

Be friendly and courteous and give coworkers and supervisors the benefit of the doubt. They have reasons for behaving the way they do, just like you.

> Norma worked hard at the farm co-op that handled feed, seed, fertilizer, and other supplies. Sometimes, Mr. Monckton, her supervisor, was short and angry: "Norma, help this customer on telephone line three." Norma was angry at the abrupt tone, but controlled an impulse to shout back. She asked a coworker what she was doing wrong. The worker said: "Don't let Mr. Monckton bother you. He always gets tense and bossy when things are real busy."

The Hard Times. At times it will seem like the only thing to do is quit. A disagreeable work assignment or an impossible new supervisor may make work seem intolerable. Don't make a hasty decision. It is important to try to work through the situation.

It is important to work through difficult situations. At the end of a hard week, there is satisfaction in having completed a job and in having done the job well.

> Norma dreaded days when deliveries were made. As a recent employee, she got the job of checking off each item as it came into the warehouse. She got careless with her figures. Mr. Monckton told her: "I know you don't like this task. But it is very important to our operation. We need to know in detail every item in our inventory. I believe new employees need this experience before being promoted. It gives them familiarity with our product line. I did this same job 15 years ago." Now Norma understood why the job was important to

the company and her. She did it more carefully and in a better spirit. When she was promoted, checking deliveries became the job of another new employee.

Serving the Community. You may find that work does not adequately meet your need to serve others. There are many community organizations that provide services to the community and to individuals. The work of these religious, civic, political, and community groups is necessary to the effective working of America's democratic system of government and way of life.

Working actively with the FFA helps prepare you for effective participation in community organizations and for work. These organizations may offer you the chance to develop your leadership abilities before you get a chance at work. There will be committee assignments and other activities that can challenge your leadership skills.

In addition to the personal satisfaction you gain from community service, the effort could show your employer that you are a responsible person who is willing to take on challenging assignments. This could be another point in your favor when the next promotion is available.

Work and Personal Growth: A Review

The workplace and community service provide much the same kind of opportunities that you find in school and in organizations like the FFA. You must continue to learn in work and community life in order to advance, to grow, and to find personal satisfaction. Along with learning new things, you will practice your skills, serve your community, and do the work that serves your own needs and those of your community and nation.

The FFA motto is appropriate for your life as you prepare to leave school:

Learning to do
Doing to learn
Earning to live
Living to serve

THINKING IT THROUGH

1. What type of job in agricultural industry should you look for?
2. How can job openings in agricultural industry be located?
3. How can you decide if a job in agricultural industry is right for you?
4. What are the steps in applying for a job in agricultural industry?
5. What should be listed on a personal data sheet?
6. How should a prospective employer be contacted?
7. What are the ways of preparing for a job interview?
8. What should be done during the interview?
9. How does the work situation relate to group dynamics?
10. What will help in getting along well with other people?
11. What are the ways of preparing for job advancement?
12. What are some ways of being of service to others?

GLOSSARY

Achievement The act of mastering something, especially through skill or hard work.

Address the chair To obtain the recognition of the chair.

Adjourn To end a meeting.

Affiliate To become a part of a group.

Agenda A list that specifies the order in which the business of a particular meeting is to be conducted.

Alternatives The various options people have available to them.

Altruism The attitude and practice of being of help or service to others.

Amend To make a change in a main motion.

Appearance How a person looks to other people.

Apply for a job To submit one's qualifications to an employer, usually on an application form provided by the business.

Authority The power or influence that accompanies a position.

Autocratic leader A person who maintains a style of leadership in which he or she uses power and position to force decisions on others.

Budget A planned approach for the wise use of time or money.

Building Our American Communities (BOAC) An FFA program designed to promote the improvement of local communities.

Body language Nonverbal information that is communicated through a person's body gestures.

Bylaws A written document that follows the constitution and outlines operational procedures for an organization.

Capacity A person's ability to make others feel he or she knows what needs to be done and can get these things done.

Chair The person in charge of a meeting.

Change process The steps an idea goes through from its origin to the time at which a person or group adopts it and puts it into practice.

Characteristic A quality or trait identified with a person, often in a specific role.

Citizenship Accepting and performing responsibilities as a citizen of a country.

Cohesive Sticking together; in a group situation, a feeling of oneness.

Committee work The activities of a smaller group of an organization asked to identify problems or needs and/or carry out activities on behalf of the total group.

Communication Sending and receiving messages, the most important aspect of human relations.

Competition A trial of skill with another person.

Compromise A settlement reached among people differing with one another.

Conflict The state in which disagreement occurs within human relationships.

Constitution A written document that outlines the purposes, membership, and framework of an organization.

Contest An event in which individuals or teams compete.

Conversation An informal oral exchange of ideas or views.

Cooperation Working and associating successfully with others.

Coordinating Directing several activities toward a common goal, need, or problem.

Coworkers The other people with whom a person works.

Criticism Questioning the actions of another.

Debate The exchange of ideas related to the motion on the floor of a meeting.

Decision A purposeful conclusion that usually leads to a course of action.

Decisive The quality of being able to make a decision and carry it through.

Delegating Assigning responsibility and the needed authority to a person or group so it can function as an extension of the group leader.

Democratic leader A person who maintains a style of leadership in which he or she encourages participation and involvement of members in making decisions.

Discrimination An act or way of treatment that often is unfair or unjust.

Discussion An oral communication involving two or more people designed to answer questions or seek solutions to a problem.

Discussion leader The person in charge of leading a discussion.

Effective listening The ability to hear and understand the message presented.

Employee The person hired to do a job.

Employer The person who hires another person.

Environment All that surrounds humankind.

Evaluating Reviewing the activities and progress of a group or organization to determine if its goals have been met.

Executive A person who executes, or carries out, the decisions of the group that he or she leads.

Extemporaneous speaking Speaking to a group without the opportunity to formally prepare one's remarks.

FFA achievement award An individual award for an FFA member which recognizes accomplishments in classwork and leadership development.

FFA ceremonies The official ceremonies adopted by the organization for opening and closing meetings and awarding degrees.

FFA Code of Ethics A statement describing the conduct expected of members.

FFA creed A statement expressing belief about leadership and agriculture.

FFA degrees Recognition of achievement for FFA members who meet specified requirements.

FFA program of activities The guide prepared by chapter members outlining their plans for chapter activities and the ways to accomplish those activities.

FFA proficiency awards Recognition for members who excel in one of several areas of agriculture.

Financing Providing money to carry out activities.

Formal group A group organized for a specific purpose and having designated leadership.

Formal leader A person who has been elected or appointed to a position of leadership.

Formal speech A speech that is organized and prepared in advance and generally given to a specific audience.

Framework The way parts are related and responsibilities assigned, as between officers and members in an organization.

Fringe Benefits Benefits such as insurance or vacation that may be provided by an employer in addition to wages.

Gesture The use of hands or other body action to aid the communication process while speaking. Gestures may communicate information without speech (See *Body language*.)

Grooming The act of taking care of one's own body and clothes.

Group Three or more people usually having common goals who engage in communication and human relations.

Group dynamics The process whereby groups function to meet the needs of their members.

Group process The procedures and activities used to involve group members in group activities.

Group roles Specialized functions that members of a group take on in order to make the group work better as a whole.

Human relations The art and science of dealing with others on a person-to-person basis.

Image What other people think of a person, how they perceive a person.

Improvement A person's advancement to a better state, helped by his or her willingness to accept criticism.

Incentive The motivation to want to accomplish a job or get something done.

Informal group A group meeting without formal organization to deal with an existing problem or need.

Informal leader A person who has influence in a group but has not been elected to a position of leadership.

Initiative Taking the lead or starting a needed activity without being instructed to do so.

Insignia The emblem or other identifying symbols of an organization.

Integral One part of a whole, as the FFA is part of the agriculture/agribusiness program.

Integrity Honesty and good morals.

Interaction The process of sharing ideas by communication between individuals or within groups.

Interview The act of meeting with another person to discuss a problem, need, or job application.

Intracurricular A planned part of the curriculum or school activities.

Introducing a speaker Presenting the speaker to a group by providing relevant information about the speaker and the topic.

Involvement The productive use of time to get the most done.

Job application A form provided by an employer that lists an applicant's qualifications, experience, and references.

Knowledge An accumulation of facts that are useful in answering problems or making decisions.

Laboratory A place to gain knowledge, skills, attitudes, and experience.

Laissez faire A style of leadership in which there is little or no direction from the leader.

Leader A person who voluntarily works with others to help them accomplish their goals or needs.

Leadership The quality of being able to guide, direct, instruct, and/or help others take responsibility and meet common needs.

Leisure time Free time in which people do what they choose.

Letter of application A letter written by a person to an employer asking to be considered for a job.

Majority One over one-half of the persons voting.

Maturity The stage of development in which a person has become an adult and can make wise decisions affecting his or her life.

Member services Activities in which an organization engages to help its members.

Minutes The official record of the meeting of an organization.

Motion A proposal that action be taken by an organization.

Motivation The drive within people that leads them to act or function.

National symbols Items, songs, etc., that have been selected as representatives of a nation's heritage.

Needs Desires that a person feels should or must be met.

Nominating committee A group of people appointed to recommend individuals to serve as officers for an organization.

Nomination The act of naming a person as a candidate for office.

Nonverbal communication Sometimes called body language, getting ideas across by expression and movement.

Obtain the floor To secure from the chair the right to speak.

Organization A group of persons structured in some manner to meet identified needs of its membership.

Parliamentary procedure Guidelines to follow in conducting a meeting to determine the will of the majority while protecting the rights of the minority.

Participation The act of getting involved, such as in participation in FFA meetings.

Persevere To continue to work at a task even though it may be difficult.

Personal data sheet A document that a job applicant prepares listing personal information which may be helpful to an employer considering whether to hire him or her.

Personal inventory A listing of a person's strong and weak points.

Personality How others view a person because of his or her characteristics.

Plurality The most votes for a nominee when there are three or more nominees.

Precedence The order in which items or motions must be considered.

Prejudice Attitudes or thoughts based on previous ideas or experiences that caused a biased or unfair view.

Presiding officer The person who calls a meeting to order and is responsible for maintaining the order according to parliamentary procedure.

Prestige Influence reflecting the esteem others have for a person based on their respect for that person.

Public speaking Communicating by speaking to a group of people in a formal situation.

Quorum The number of members that must be present for an organization to legally transact business.

Recognition Approval and support, such as is shown with an award or honor.

Relationship The manner in which people or things are connected to one another.

Respect Being favored or viewed positively by others.

Responsibility The quality of being accountable for certain actions or activities.

Salutation The introductory words of a letter.

Second To endorse a motion before considering it further.

Security A sense of well-being and/or protection.

Self-conduct The way an individual behaves.

Self-confidence Belief in one's own worth and abilities.

Self-employed The term that describes a person who owns or operates his or her own business.

Social behavior The ways in which people conduct themselves among others.

Social skills The ability to get along with other people.

Specialize To restrict oneself to studying a special activity.

Status How others view a person's standing in a group.

Stereotype A blanket judgment about a class of people that is based on prejudice.

Structure The relationship of parts, as between officers and members in an organization.

Supervisor A person designated by a business whose job is to see that others do their jobs correctly.

Survival The act of maintaining existence.

Values The attitudes of people that relate to the things they feel are most important.

INDEX